D0810176

The Virus That Ate Cannibals

THE VIRUS THAT

Macmillan Publishing Co., Inc.

NEW YORK

CAROL ERON
ATE CANNIBALS

Copyright © 1981 by Carol Eron

All rights reserved. No part of this book may be reproduced or transmitted in any form or by any means, electronic or mechanical, including photocopying, recording or by any information storage and retrieval system, without permission in writing from the Publisher.

Macmillan Publishing Co., Inc.
866 Third Avenue, New York, N.Y. 10022
Collier Macmillan Canada, Inc.

Library of Congress Cataloging in Publication Data

Eron, Carol.
     The virus that ate cannibals.

     "A portion of this book originally appeared in different form in The Washington Post, March 7, 1977"—copyright page.
     Bibliography: p.
     Includes index.
     1. Virology—Popular works.   I. Title.
[DNLM: 1. Virus diseases.   2. Viruses.   QW 160
E71v]
QR364.E76   616'.0194   81-3732
ISBN 0-02-536250-X   AACR2

10   9   8   7   6   5   4

Printed in the United States of America

A portion of this book originally appeared in different form in *The Washington Post*, March 7, 1977.

For LARRY, ETHAN, and LUCY

vi·rus \ ′virəs \ *n* -ES [L, slimy liquid, poison, stench]

Contents

Thank You

MANY, MANY PEOPLE helped me as I worked on this book. The editors of *The Washington Post* first provided encouragement by publishing a portion of it.

For innumerable hours of interviews and conversations, and for their ideas and criticisms, all most generously given, I especially thank Sir Christopher Andrewes, Dr. David Baltimore, Dr. Robert Buchanan, Dr. H.R. Dion, Dr. Frank Ennis, Dr. George Galasso, Dr. Clarence Gibbs, Elizabeth Theiler Martin, Dr. Albert Sabin, Dr. James Shannon, Dr. Hugh H. Smith, Sir Charles Stuart-Harris, Dr. Howard Temin, and Dr. Harriet Zuckerman.

Dr. James Dahlberg, Dr. Jack Gwaltney, Dr. Elsebet Lund, and Dr. Sylvia Reed were among those who reviewed portions of the manuscript. Lady Andrewes offered warm hospitality on cold November days.

The staff of the National Library of Medicine has endured

me for years, and excellent assistance also came from the staffs of the National Institutes of Health, the Columbia University Oral History Collection, the Rockefeller Archive Center, the Walter Reed Army Institute of Pathology, and the Countway Library of Medicine at Harvard Medical School.

For their enthusiastic support, I am very grateful to Dr. Carole Horn, Brigitte Weeks, William McPherson, Theron Raines, Robert Levine, and two fine editors, Marion Wheeler and Joyce Jack.

Last, I am grateful to Larry, my husband, who provided expert advice and whole-hearted encouragement from start to finish.

1

"Obscure, If Not Positively Unnatural"

IMAGINE A CREATURE that is nearly invulnerable and so small it cannot be seen under an ordinary laboratory microscope, although it possesses powers of destruction that rival our own most awesome weapons. A speck of chemical matter much smaller than a single cell, smaller even than protozoa or bacteria; a creature much less than a millionth of an inch in size, yet so tough it can survive a temperature of 200 degrees below freezing, and a force 100,000 times greater than gravity cannot crush it.

Drifting this way and that, driven by wind or water, riding about inside an animal, insect, or bird, it is quite at home in the sea, the earth, a green plant, on a doorknob, a fingertip, or the page of a book. It can endure for hundreds of years under hostile conditions; returned to hospitable terrain, it springs to virulent life.

Virus. The word rings of malevolence, of nature in one of its most mysterious and sometimes poisonous forms. Viruses speak of devastating epidemics, of smallpox, yellow fever, influenza, polio. They mean rabies, herpes, hepatitis, pneumonia, and a host of exotic lethal afflictions—Lassa fever, encephalitis, dengue. They mean milder annoyances—measles, mumps, chickenpox, warts, cold sores, the common cold. Viruses may also help to cause certain cancers, and lately, they are being blamed for diabetes and a vast array of unexplained neurologic diseases.

In many ways, viruses have proved a far more formidable foe than the microorganisms that produced the Black Death and the other great plagues of the Middle Ages. Viruses have proved impervious to antibiotics and resistant to most sanitation measures, vaccines, and other attempts to control them. They are to blame for almost eighty percent of all acute illnesses in the United States each year, with influenza and other respiratory illnesses alone responsible for approximately a third of all visits made to physicians. The herpes virus causes a genital infection afflicting millions of Americans, for which there is no known cure. In extreme form, the herpes virus deforms and kills infants. Even the lowly cold viruses take their toll, causing the loss in this country of 40 million workdays a year.

Viruses possess the strength of numbers. More than one hundred different families of viruses are recognized, each with countless members. In the case of the cold viruses, this means there are too many active strains to be incorporated into a vaccine. Even if a huge vaccine could somehow be made, these viruses have a further advantage in that the immunity we develop to them does not last very long—a few years at most. And for many colds, no virus can be found. As a result, the cold viruses proliferate far and wide, causing a billion colds a year in the United States alone and an expenditure of $750 million annually on cold and cough remedies.

But there is another side to the story. Viruses are not simply agents of misery and destruction; the fact is that we are surrounded by viruses, inside and out, and most of the time they do us no harm at all. We go about our business, and they go about theirs.

And what might their business be? It seems to be genetic engineering—the manipulation of pieces of genetic information, strings of genes, between one living organism and another. It may be that when we are forced to take notice of viruses, when they make us ill or devour our crops or our livestock, they have just made a mistake. And there is at least one virus known to create a work of beauty—the tulip break virus, which streaks hybrid tulips with color. So prized were the virus-infested bulbs in seventeenth-century Holland that wild speculation broke out—a phenomenon known as tulipomania—and government interference was necessary to restore order.

You might be consoled to know, too, that for all their ferocity, viruses have a great weakness: they need us! Unlike bacteria, which reproduce by themselves, merely by dividing in two, viruses can function and multiply only inside living cells. Until a virus has the good fortune to enter a cell—be it human, animal, vegetable, or bacterial—it is for all purposes dead.

Far more elemental than bacteria or human cells, viruses contain only one type of nucleic acid, either RNA or DNA. They also lack the ability to produce energy and manufacture proteins. Therefore, viruses exploit cells, using their enzymes to produce energy, proteins, and more nucleic acid.

Beneath a virus's protein coat of armor lies a bit of transparent jelly that is its nucleic acid—DNA or RNA—and holds the complete genetic instructions, the master plan for manufacturing more viruses. But the virus does not have its own factory or raw materials for reproduction. These are supplied, under coercion, by living cells.

When a drifting virus bumps into a cell for which it has a special affinity, the attraction is irresistible. The virus clings to the cell's surface. Feeling the tickle, the cell then puckers up, surrounds the virus, and gulps it in. Some less patient viruses, such as influenza, blast a hole in the cell membrane and fire their contents inside.

Having admitted a stranger, the cell errs a second time by dissolving the visitor's coat with an enzyme intended to kill—unleashing its deadly essence: fine, long, intricately coiled strands of RNA or DNA, the genetic command material, which acts as a template for the virus's reproduction. Enslaving the cell to its own ends, the virus issues orders, shutting down the cell's vital manufacturing processes, and starting production of materials that will form viral progeny.

Twenty different amino acids, the building blocks of the proteins necessary for life, are then linked together in millions of different permutations, forming the enzymes that direct various stages of the reproductive process. Other enzymes are manufactured to bore a hole in the cell membrane once the young viruses are ready to escape. Then mortally wounded, the cell usually dies. Depending on the kinds of cells destroyed, and how many, this loss may do us no harm at all, or it may do us in.

Different from bacteria, which attack from outside the cell, a virus is protected by the cell itself during its vulnerable reproductive stage, when the virus is stripped of its armor. Some viruses go so far as to integrate themselves into a cell's very genes, where they are in a position to make fundamental changes in our genetic make-up. But in either case, the virus is quite safe, since a drug that could kill the virus would be likely to kill the cell as well.

Some viruses are chameleons—another great strategic advantage. Influenza viruses, for example, can transform themselves to such a degree from year to year that as far as our immune systems can tell, every time they strike, they seem to be brand-new viruses. In the fall of 1918, when such a new strain emerged, known as the Spanish flu, it spread like wildfire throughout the world, killing 20 million people in a single year. No war, famine, or disease, not even bubonic plague, has taken so many lives in such a short time.

A huge reservoir of influenza viruses has been found in migratory birds, swine, and other animals, which may be a source of the new epidemic strains. The viruses are constantly being passed around; mallard ducks, for example, carry influenza viruses which they spread to turkeys, and chickens in Scotland are known to carry an influenza virus that has been related only to a virus of migratory terns from South Africa. Such evidence has generated the theory that human influenza viruses, passing through animals, combine and exchange genetic information with animal strains and then at times reemerge metamorphosed to produce epidemics in us.

No one knows where viruses began. They may have degenerated from some larger disease-causing organism that once lived independently. Or they may be escapees from cells—genes that ran away and evolved to survive on their own.

Some viral diseases have existed since the world began, and viruses have been fellow travelers with us all along. One of the earliest known victims of a virus was the Egyptian king Ramses V, whose pocked, mummified face reflects the smallpox of which he died three thousand years ago. Other viral diseases have grown much more troublesome since then, occurring more often and more severely. The word *influenza* was introduced in Italy to describe an epidemic early in the fifteenth century, when such mass tragedies were attributed to the influence of the stars; the first report of a continent-spanning influenza epidemic dates as recently as the late sixteenth century.

Earlier in the sixteenth century smallpox took the lives of several million Aztec and Inca Indians in Mexico and South America in only a few years and probably enabled the vastly outnumbered Spanish to conquer them. In the seventeenth and eighteenth centuries smallpox emerged as the most devastating disease to strike Europe and the New World in history, and in the eighteenth century alone it killed 60 million people.

There is evidence that the Spanish used smallpox for germ warfare, intentionally introducing it to some groups of hostile Indians. But the great burgeoning of viral epidemics was probably due to such developments as the growth of crowded cities and new patterns in travel and trade rather than to intentional efforts to disseminate viruses, better record keeping, or sudden major transformations in the organism. Of course, no one can say for certain that viruses did not undergo fundamental changes in the centuries immediately before they came to our attention, but it doesn't seem likely.

That such a creature as a virus existed was unsuspected for several hundred years after bacteria were first sighted under a microscope by Anton van Leeuwenhoek in 1683. By the late nineteenth century many kinds of bacteria had been found and identified, but researchers began to remark on certain curiosities that seemed at odds with the new scientific knowledge. A Dutch botanist, Martinus Beijerinck, who was studying a disease of tobacco plants, noticed that something from the pressed juice of diseased plants was sneaking through his finest porcelain filters' used for straining bacteria, and it was creating chaos with his experiments.

This filtered fluid, Beijerinck observed, had a bizarre character: first it appeared in his laboratory, then it disappeared, and just when he thought it was gone for good, it reappeared, as infectious as ever. He was forced to propose that this "contagious living fluid," unlike bacteria, which would grow even in a piece of old cheese, might grow only in living cells. But he could not quite believe this; the idea struck him as "obscure, if not positively unnatural." In 1898, Beijerinck wrote: "It seems to me that reproduction or growth of a dissolved particle is not absolutely unthinkable, but it is very hard to accept." He called the mysterious new substance a *virus*, which in Latin denotes a slimy liquid, poison, or noxious stench.

Those who have set out to battle viruses have had a devilish time. Almost invariably, tactics that succeeded once failed to work again. After an English country doctor, Edward Jenner, proved in 1796 that smallpox could be prevented by inoculations of a much milder disease, cowpox, ninety years passed before another great victory against viruses. In 1885, Louis Pasteur applied techniques he had used to fight bacterial diseases in cattle to weaken the lethal rabies virus, and he made his legendary vaccine for rabies. Pasteur called it a vaccine, meaning "of the cow," in honor of Jenner, although the substance came from the saliva of a mad dog. Like Jenner, Pasteur did not know what had caused the disease his vaccine cured, or exactly how he had achieved success.

Both vaccines contained "live" viruses, which though weaker than the most deadly smallpox and rabies viruses, were still similar enough to them to stimulate the production of antibodies—special proteins that can identify the foreign markings, or antigens, of an invading microbe and destroy it. Later, effective vaccines were made with "killed" viruses—viruses inactivated with heat or chemicals, while retaining their antibody-producing properties. But it was an arduous process of trial and sometimes tragic errors; and the early vaccine makers never could have fully anticipated the enormous difficulties and traps that lay ahead.

With two splendid triumphs in hand by the end of the nineteenth century, it was expected, reasonably enough, that other viruses—whatever they were—would shortly go the way of smallpox and rabies.

Knowledge grew tremendously in the following years. As early as 1908, viruses were first linked to cancer—in chickens. By 1935 a virus was purified and its composition determined, thus settling the great question of what a virus is. With the invention in Germany of the electron microscope, viruses were finally seen—in 1939.

But it was fifty years after rabies was tamed and conquered before another great medical victory was won: the resounding defeat of yellow fever. Methods of growing viruses had to be developed in order to study the viruses and then to make virus vaccines. At first, viruses were cultured in animals, a slow laborious process that yielded unreliable results. Eventually, it was learned how to cultivate viruses in bits of animal tissue, first on glass slides—this was accomplished by Dr. Edna Steinhardt at Columbia University in 1913—then in dishes, tubes, and larger flasks. By passing viruses from one flask full of animal cells to another, generations of viruses could be produced exponentially: one virus infecting one cell produces a hundred viruses, which are released from the cell and infect a hundred more cells, and so on. In this way billions upon billions of viruses could be produced. But bacteria also thrived in the rich nutrients used for animal tissue culturing, and not until antibiotics were developed did the technique really advance.

In the case of yellow fever, and succeeding cases, other critical clues kept turning up in the most unpredictable places: in a jungle, the ocean, a ferret, a sheep, a monkey, an ordinary mosquito. These clues were scattered from Senegal, to Iceland, to Italy, to Florida, from the English countryside to the highlands of New Guinea. They were unassociated and uncodified, unseen until the time was ripe and minds were prepared to see and comprehend them and fit them into place. Luck played its part in the process, of course, as did genius, determination, golden errors, and mostly, drudging hard work.

Behind each of the six contemporary campaigns described in this book are countless others, forming the base on which one crowning achievement is built. In offering these six accounts I have, as any chronicler of historical events does, chosen those tales that could be told: those for which the records have been preserved and which people were willing and able to recall. Neither is this chronicle intended as complete. For despite all that has been achieved, the fight against viruses is far from over. It is just beginning.

2
The Sting of Death

RETURNING HOME FROM A TRIP abroad in the spring of 1928, Dr. Andrew Watson Sellards carried a small, tightly sealed Thermos bottle. During the long voyage from French West Africa to Boston via London, he kept a close watch over it and fussed constantly with the wrappings. Within, packed in ice and sea salt, was a glass tube containing a bit of evil-looking brownish substance that undoubtedly would have caused a customs inspector apprised of its exact nature to faint.

It was nothing more or less than the frozen liver of a monkey, poisoned with one of the most dreaded plagues known to humanity: yellow fever. Dr. Sellards had obtained this particular liver after months abroad and at great expense. He was anxious that nothing happen to it during the journey, so he continually replaced the ice in which it was packed even before the ice began to melt.

The poisonous liver was the product of Dr. Sellards's fruitful collaboration with two French researchers at the Institut Pasteur in Dakar, French West Africa. The previous October, of 1927, Dr. Sellards happened to read of the death of a yellow-fever researcher, Dr. Adrian Stokes, in the *British Medical Journal*. The notice included excerpts of letters Stokes had written concerning his work in Africa, and, as he perused the information, Dr. Sellards made up his mind to go to West Africa and conduct research along the same lines. On November 24, 1927, he arrived at the Institut Pasteur with a dozen monkeys in tow and a cage full of ordinary domestic mosquitoes.

After exposing sixteen mosquitoes to a man ill with yellow fever, Sellards and the French scientists set the insects loose to feed on the monkeys. With great excitement they observed that the bitten monkeys caught yellow fever and died. This method of killing a monkey with yellow fever was, as a matter of fact, a great feat. It had first been accomplished by Stokes and his colleagues six months before, and it represented the first advance against yellow fever in many years.

Until yellow fever was transmitted to monkeys, it had been thought that only humans were susceptible to the terrifying illness. The discovery that a laboratory animal was also susceptible meant that a serious assault on the disease could begin.

Dr. Sellards planned to take the lead. With his monkey liver and a few of the deadly mosquitoes, which he also brought home to Boston, he would be among the first in the United States to experiment with yellow fever, and he was extremely pleased. It had been known ever since 1901 that mosquitoes were the culprits that spread yellow fever, but no one knew what it was they were spreading. Dr. Sellards intended to find out.

For several centuries previously, yellow fever had rampaged the eastern and southern coasts of the United States, keeping citizens in a state of continual fear. It had first traveled west on slave ships in the seventeenth century; and in combination with other imported infections, yellow fever destroyed the native Indians of the Caribbean. The Africans who replaced them were more resistant to the disease.

Because it killed so quickly and horribly it was the most

dreaded of all plagues. More feared than even malaria, it was called yellow jack by British soldiers after the yellow quarantine flags flown in ports and on ships to warn of its presence. Ships went down in storms rather than enter ports where the yellow jack flew.

By the time the first victims began turning lemon yellow and the terrifying black vomit started, it was often too late to escape the quarantines that were swiftly imposed. Those who fled quarantine risked being shot as they approached another town or farmhouse.

The cause of the disease was elusive. In 1793 the most eminent physician in Philadelphia, Dr. Benjamin Rush, argued that yellow fever, which had swept the city and reduced it to near chaos, originated in the "putrid exhalations" from some spoiled coffee that had been dumped on a city wharf. "It is no new thing for the effluvia of putrid vegetables to produce malignant fevers," Dr. Rush wrote in his *Account of the Bilious Remitting Yellow Fever*. "Cabbage, onions, black pepper, and even the mild potatoe, when in a state of putrefaction, have all been the remote causes of malignant fevers." Sulphur was burned in the city streets and the dock was scoured, but neither measure halted the disease. In one of the increasingly frantic attempts that were made to clear the putrid air, soldiers fired cannons up and down the streets, but this was to no avail, and the practice was abandoned.

Within a week fear enveloped the city. People began avoiding one another, and if they happened to meet they no longer shook hands. They stayed indoors as much as possible. When going out they held handkerchiefs sprinkled with vinegar and camphor to their noses. Physicians recommended that standing or sitting in the sun or evening air be strictly avoided as well as trotting vigorously on horseback and intemperance, though drinking wine and cider in moderation was highly recommended.

In two months the epidemic killed one of every ten Philadelphians. Those who were able fled the city. Working from dawn till dark, Dr. Rush was forced to turn away more patients

than he could see, and his carriage was waylaid by people pleading for help. As the epidemic worsened, all his apprentices fell
ill and died, and before it was over he had lost his father and
sister, too. Ill and exhausted himself, Dr. Rush continued to minister in his parlor to as many as one hundred patients a day, although the situation appeared utterly hopeless.

The death toll climbed daily, from three to five, twelve, fifteen,
seventeen, and more. When the College of Physicians met to consider the burgeoning disaster, they agreed that the incessant toll-
ing of the church bells for the dead was having an unhealthful
influence and should be stopped immediately. Equally dangerous,
Dr. Rush believed, were strong emotions, such as a "sudden
paroxysm of fear" or sudden relief from anxiety, which could
excite the contagion into action in the body. He had drawn this
conclusion from his observation that the families of yellow fever
patients who recovered were as likely to catch the fever as the
families of those who died from the disease. As for the actual
poisons, Dr. Rush admitted, "by what process of nature they are
generated, few pretend to know; and though some may fancy
they know them, yet it is beyond the conjecture of most."

On entering a patient's room, the first thing that struck Rush
was the victim's countenance. "It was as much unlike that which
is exhibited in the common bilious fever as the face of a wild is
unlike the face of a mild domestic animal. The eyes were sad,
watery, and so inflamed in some cases as to resemble two balls of
fire. Sometimes they had a most brilliant or ferocious appearance.
The face was suffused with blood, or of a dusky colour, and the
whole countenance was downcast and clouded . . . sighing attended in almost every case."

His usual treatment consisted of vigorous purges and bleedings
(which, with all that infected blood splashing about, must have
done more to decimate the ranks of the healers than any other
treatment). "In determining the quantity of blood to be drawn
I was governed by the state of the pulse, and by the temperature
of the weather," he wrote. In warm weather a small amount was

drawn, in cooler weather, much more. But all his remedies failed, except possibly, he thought, for wrapping patients in hot blankets soaked with vinegar.

Many a priest and city official fled their responsibilities in terror, and so many officials were away, sick, or dead that the government broke down. Ships remained docked for weeks because their crews were ill, preventing incoming ships from docking, had they even wanted to. Businesses collapsed. The city's clocks failed, and there was no one to fix them. All the schools closed; orphans wandered the streets. Then, after holding Philadelphia in terror for several months and killing five thousand people, the plague departed, as abruptly as it had come. People observed with relief that the frost had seemed to kill it. But in a few years yellow fever returned to Philadelphia, and during the next century it became a terrible menace in ports from Boston to New Orleans.

There were many theories about its cause; miasmas, filth, fomites (contaminated objects such as clothing and bedding), and meteorologic conditions were blamed. Some believed that yellow fever was imported from the West Indies and that strict quarantine measures could prevent it. But the epidemics continued, and by the early nineteenth century, the contagionist view fell into disfavor. Quarantine regulations were regarded as superstitions and were relaxed. The miasmatic theory—which proposed that heat in combination with damp animal or vegetable matter created putrid exhalations—became popular and led to sewer construction, waste removal, and other sanitary measures. Still, the epidemics worsened.

In Haiti, yellow fever routed Napoleon's troops, killing most of his 25,000 men in less than a month, and helped convince him of the wisdom of retreating from Louisiana. Yellow fever devastated New Orleans in the 1850s, killing 20,000 people in ten years. In 1878 an epidemic swept up the Mississippi River from Baton Rouge to St. Louis, taking 13,000 lives and causing business losses in the millions. Twenty thousand workers died of the disease in Panama between 1881 and 1889, and yellow fever, together with malaria, forced the French to abandon construction of the canal. Along with malaria, yellow fever made West Africa the white man's grave.

Yellow fever took a toll on American troops during the Spanish-American War, thus threatening to halt American imperial ambitions in the Caribbean. After the war a U.S. Army commission headed by Major Walter Reed was dispatched to Cuba to study yellow fever. The commission arrived in Cuba in the summer of 1900. At the time, yellow fever was attributed to filth, and the army's chief sanitation officer launched a massive cleanup campaign in Havana. This, too, failed to stop the disease.

A physician in Havana, Dr. Carlos J. Finlay, had been claiming for years that "an agent in the air"—specifically, a mosquito—transported yellow fever. Dr. Finlay had not been able to convince anyone else of the validity of his claim, however, and had been dismissed. But since the cleanup campaign had not halted the disease, and in view of the recent important discovery that an insect carried the malaria parasite, the commission set to work with the larvae of some mosquitoes given to them by Dr. Finlay.

Because at that time no animal was known to contract yellow fever, the commission had to use humans for their experiments. They did not lack volunteers. Few soldiers thought yellow fever could be caught from an insect, and poor Spanish immigrants eagerly volunteered for the experiments in exchange for several hundred dollars.

In tightly screened cottages constructed for the purpose, one group of volunteers was exposed to mosquitoes that had recently fed on yellow fever victims. Other volunteers slept on the blood-blackened bedding of victims and wore their clothing. Only those who were bitten by the mosquitoes contracted yellow fever.

Many of these volunteers survived; some did not. Three months after the work began, Jesse Lazear, an entomologist who was a member of Walter Reed's commission, died from the bite of an infected mosquito. In the case of the young American nurse

Clara Maass, who twice volunteered to be bitten, the second bite proved fatal. "No soldier in the late war placed his life in peril for better reasons than those which prompted this faithful nurse to risk hers," the *New York Times* wrote at her death. By then the commission had succeeded in proving beyond doubt that mosquitoes spread yellow fever, so the experiments ended.

Reed studied the filtered blood of victims and decided that since neither he nor anyone else could discern anything unusual yellow fever must be caused by a creature so small it was "ultra-microscopic."

Following the Reed commission's discovery in 1901, squads of mosquito hunters set out to destroy the insect. By good luck, the particular mosquito that spread yellow fever, *Aedes aegypti*, was a highly domesticated species that breeds in small, still pools of water and prefers artificial containers to those with a sandy or muddy bottom. Under the direction of Major William Gorgas of the U.S. Army, mosquito inspectors climbed ladders and poked into rainspouts, drains, sewers, cisterns, water casks, jugs, and holy-water fonts all over Havana and made sure they were covered or else, after fair warning, destroyed them.

It was the most intensive cleanup campaign ever undertaken, and as a result the disease seemed to vanish. Less than a year later Havana was free of yellow fever for the first time in 150 years. After similar campaigns, yellow fever disappeared from the United States and Panama, and construction of the canal proceeded. By 1925, yellow fever also disappeared from Mexico and Central America, and a cleanup campaign was launched in Brazil.

In 1919, eight years before Dr. Sellards set off for West Africa, the great bacteriologist Hideyo Noguchi had announced that he had found the agent responsible for yellow fever—spiral bacteria. Noguchi's spirochete did produce some of the symptoms of yellow fever in guinea pigs, but the ailment was not easily spread among the animals by mosquitoes. This was in odd contradiction to yellow fever in humans, which spread all too well. There were suspicions about Noguchi's discovery, and the confusion and debate stirred by his claims persisted from 1919 until 1927, when Dr. Adrian Stokes, an English pathologist, and his colleagues first transmitted yellow fever to monkeys. They also demonstrated that the disease was caused by a "filterable virus." By this they meant that the organism was smaller than ordinary bacteria, some kind of minute infectious creature so small it was capable of sneaking through the fine porcelain filters used to screen out bacteria. It reminded them of the filterable fluid Louis Pasteur had first come across in the 1880s that caused rabies. But more than that no one knew. "You shall know a virus by its deeds," researchers remarked for decades after that. It was all that could safely be said on the subject.

Dr. Sellards, a professor at the Harvard School of Tropical Medicine, held the prevailing view of the era, that yellow fever was probably caused by some form of bacteria, if not necessarily Noguchi's spirochete. His laboratory assistant, a soft-spoken, usually reticent young South African named Max Theiler, disagreed. Theiler believed that the newest ideas were correct, that the organism might be altogether different from bacteria—a form of life that couldn't reproduce almost anywhere, like bacteria, but a finicky creature that thrived only in living cells.

In another laboratory at Harvard, the herpes simplex virus had been injected directly into mouse brains, and Max Theiler wondered what would happen if he did the same with the yellow fever Dr. Sellards had just brought from Africa. Although researchers had tried injecting contagious yellow fever fluid into the bellies of mice and beneath their skin, without success, no one had thought to try mouse brains.

Dr. Sellards was not interested in experimenting with yellow fever and mouse brains, and Theiler resolved to proceed quietly on his own. When he first came to work in the laboratory, the professor had treated him like a son, but Theiler thought he must have displeased Dr. Sellards somehow, for lately he had become irritable and contrary.

For many reasons, Theiler hesitated. He was indebted to Dr. Sellards and always would be. After completing his medical training, Theiler had decided he was not cut out to be a doctor. He had found the practice of medicine depressing, and surgery had not interested him; it seemed mechanical.

In short, he hadn't known what he wanted to do, which was an embarrassing predicament for the son of Sir Arnold Theiler. As a young, unknown Swiss veterinarian, his father had arrived in the wilds of Pretoria, South Africa, in the 1890s and had gone on to establish a leading veterinary research institute. His father had made many important contributions to the development of veterinary science and was something of a legend. As a young man, Sir Arnold had lost a hand cutting fodder, and the boys who tended his research animals believed that his wooden hand in its black glove was possessed with magic. His own children were also very much in awe of him, although Max, the youngest of four children, was mischievous and frequently defied his father. Unusually small for his age—his friends called him "Tick"—Max was determined to do everything the other boys did.

In school, much to Sir Arnold's displeasure, Max was only average—and determinedly so. "The prevailing view among the boys was to do as little study as possible to get by," he explained once. "The hard-working boy—or swat—was looked down on. In subjects which I really liked, I studied far more than was required, but always taking care that I did not appear to be a swat." His main distinction was that he could walk up the school steps on his hands. Life at the English-style school was Spartan, with cold showers every morning, canings for misbehavior, and bad food, particularly the oatmeal, which he said "was sometimes highly unpalatable due to the presence of maggots."

But home was only eight miles away, so he fastened his laundry bag to his bicycle and pedaled home every Friday afternoon.

There were family trips to the zoo on Sundays, which meant an elephant ride, a look at the crocodile donated by Sir Arnold, and eating ice cream in a shaded outdoor cafe to the music of a British regimental band.

When Max became interested in insects, his father gave him entomological equipment, though he was too busy to accompany his son on outings. Sir Arnold did take an active interest in his children's education, altogether too much interest for their comfort, and Max said more than once, "I took the first boat out after the First World War." In 1919 he arrived in London, where he studied medicine.

Hoping that tropical medicine might prove more interesting than general practice, Max enrolled in a short course at the London School of Tropical Medicine and Hygiene. There he heard from another student about a job at Harvard. In October 1922, aged twenty-three, Theiler arrived in the United States and met Dr. Sellards. The job was his.

Theiler learned research methods from Dr. Sellards, who impressed upon him the importance of experimental controls. Dr. Sellards also got Theiler interested in yellow fever. Theiler had originally thought his stay at Harvard would be a short one, but he liked research so much that he remained, despite his determination not to be a scientist like Sir Arnold. "If this is the scientific life," he often told himself as he observed his striving, politically ambitious father in action, "that's just what I'm not going to have."

For a time he and Dr. Sellards got along splendidly. Dr. Sellards could be charming and fatherly; but he could also be cantankerous, and his list of dislikes was long. As time passed, Theiler sensed that he was in danger of being added to the list.

A bachelor living at the Harvard Club, Dr. Sellards led a rather lonely existence, and he had not seemed pleased when Theiler married a Harvard Medical School student, Lillian Graham, who worked in the department. While that probably did not explain Dr. Sellards's displeasure with him, Theiler believed that it probably hadn't helped him either.

In the summer of 1929, while Dr. Sellards was away on vacation, Theiler rounded up some mice, "any odd mice that I could lay my hands on." In his position he couldn't dream of experimenting with monkeys; they were scarce, expensive to purchase, and even more expensive to maintain. Theiler's small stipend was barely enough to live on; it did not include any funds for research materials. But if yellow fever could be given to mice, which cost only pennies apiece, then science would have a fine laboratory tool, and even a lowly laboratory assistant could take a crack at discovering the cause of yellow fever. He discussed the idea with Lillian, who volunteered to help him.

Theiler reminded himself, as he made plans for injecting yellow fever germs into mouse brains, that he would have to be extremely cautious. Even in the laboratory, those who were most careful had found yellow fever perilous to work with, and it had dispatched a good many of its presumed conquerors. Extremely contagious, it came on suddenly, and there was no treatment or cure. It was often fatal. The initial symptoms resemble influenza: aches, pains, a high fever, and sometimes nausea and vomiting. The tongue often turns scarlet. In about three days the fever drops and the patient seems to be recovering. This lasts for hours or days. In a mild case the patient recovers completely, but in more serious cases a final period of "intoxication" commences: the fever rises, jaundice sets in, and the infamous black vomit—blood blackened by stomach acids—appears as the liver, heart, kidneys, blood vessels, and brain are attacked.

The first researcher to fall victim to yellow fever had been Jesse Lazear, a member of Walter Reed's commission. The following year, a British doctor investigating yellow fever in Brazil contracted the disease and died. Then Howard Cross, a Rockefeller Institute scientist working in Mexico, succumbed shortly after drawing blood from a yellow fever patient.

Next to die was Dr. Stokes, who had worked with the Rockefeller Foundation Yellow Fever Commission in Africa. Stokes, who was forty when he died, didn't wear gloves in the laboratory, and his associates thought he had contracted the fever through a small, partially healed sore on his finger. Before he died, Stokes insisted that he be bitten by two hundred mosquitoes to further yellow fever research, and this was done.

In 1928, Hideyo Noguchi, who had gone to Africa to try to prove his discredited spirochete theory, died from yellow fever. Whether his death was accidental or a suicide was unknown. After Noguchi died, *Aedes aegypti* mosquitoes were found buzzing about his laboratory. Ten days later the physician who had drawn blood from Noguchi at the start of his illness, Dr. William Young of the British Medical Research Unit in Accra, was also dead of yellow fever.

The latest casualty was Dr. Paul Lewis of the Rockefeller Institute, who died at the end of June 1929. Yellow fever killed him in just four days.

Max Theiler cautiously pulled on a pair of heavy leather gloves several weeks later and removed Dr. Sellards's poisonous monkey liver from the icebox. He cut away a tiny slice, ground it up with mortar and pestle, and suspended it in fluid. Lillian held each mouse while he injected the fluid into their brains with fine, sterile needles. Six mice were injected. Then they waited.

In a week all six mice were dead, so Theiler set to cutting them up, hunting for signs of yellow fever. He examined their livers, which yellow fever attacks first, then their hearts, kidneys, and brains, but he could detect no trace of yellow fever. The mice were not even jaundiced. The only point of interest was an inflammation of their brains, and Theiler assumed they must have died of encephalitis. It was apparent that an error had occurred in the experiment, so Theiler repeated the procedure.

The second batch of mice also died of encephalitis. Not knowing what else to do, Theiler took the yellow fever viruses from one of the dead mice and injected them into the brain of a healthy mouse. When that mouse died of encephalitis, he took the yellow fever viruses from its brain and injected a third mouse, which also died of encephalitis. Theiler was distressed. The yellow fever germs from Dakar always caused fatal yellow fever in monkeys. Certainly contamination could not explain the encephalitis: he had been too careful, and it had happened once too often. Another explanation crossed his mind, but he knew it was absurd, so he dismissed it.

When Dr. Sellards returned from vacation and learned of Theiler's research, he gladly gave him a rhesus monkey, which is much higher on the phylogenetic scale, and much closer in its immune reactions to humans, so the muddle—an interesting one, Dr. Sellards admitted—could be cleared up. Theiler injected the monkey's belly with yellow fever viruses that he had previously injected into the brains of three mice. The monkey died of typical yellow fever.

Theiler decided that was not altogether bad: it meant, at least, that the yellow fever germs had not disappeared on their journeys through the mouse brains, and it probably meant much more. He was sure now that he was onto something big, but he tried to stifle his excitement. To find out he needed more monkeys.

Dr. Sellards had made it clear that he would not supply more, so Theiler scrounged up two monkeys on his own. After collecting the yellow fever viruses from the dead monkey, he injected them into several mice and then into one of these monkeys. He watched in fascination as the monkey grew feverish and then returned to bursting good health.

After injecting the virus into several dozen more mice brains, he injected the second monkey—which did not become even slightly sick! Yet the ferocity of the virus Dr. Sellards had brought back from Dakar was *infamous*—a mere drop of it placed beneath the skin was deadly. He decided there could be only one explanation: inside the brains of these mice, the mild

yellow fever virus must have changed, grown *tame!* Only Louis
Pasteur had ever before tamed a virus when he passed the rabies
germs again and again through rabbits' brains until they gradually
became less virulent. The only other vaccine for a viral disease,
smallpox, had been a gift of nature, a milder form of the disease
found in cows. It was presumptuous to think that he, Tick
Theiler, had done what Pasteur had done, but it did look as if he
had! Mustering as much nonchalance as he could, he mentioned
the possibility to his professor.

Dr. Sellards and others quickly pointed out that this was wish-
ful thinking, not science. They said he had no business claiming
that what he had taken out of the mice and injected into the
monkeys was yellow fever, since his only evidence was one dead
monkey! Theiler knew they were right, but he kept thinking
about the two monkeys that had survived his injections. . . . He
knew he had something; the trick would be to prove it.

Soon after, Theiler awoke one morning feeling miserable; he
wondered if he was coming down with the flu. As the day progressed, his temperature rose and his whole body began to ache.
He felt ill enough to go to a hospital.

Doctors at the Peter Bent Brigham Hospital did not know
what he had. In a week he was better; the fever and aches were
gone. The eventual diagnosis of mild yellow fever made Theiler
feel better still. He regarded it as another sign that yellow fever,
and not just encephalitis, was loose in his laboratory. Back at
work, he realized how he could prove it.

Shortly before his death, Adrian Stokes developed a yellow
fever immunity test: the subject's blood sample as well as yellow
fever viruses were injected into a monkey. If the blood donor
were immune, the monkey was protected and did not contract
the disease; otherwise the monkey died of yellow fever. Instead
of monkeys, Theiler decided to use mice for his experiments.

The results were just as he had expected. When Theiler gave
mice tamed yellow fever viruses and serum from recovered yel-

low fever patients—including some of his own blood—it protected them against encephalitis. Given serum from nonimmune donors along with the tamed yellow fever virus, the mice died of encephalitis. To be sure, Theiler repeated all his experiments from the beginning. He took fresh, active yellow fever viruses and passed them from monkeys to mice and back to monkeys again. The monkeys lived. *The virus had lost its virulence!*

Theiler wrote a report on the amazing transformation of the yellow fever virus and suggested that the docile viruses might eventually be used as a vaccine. Before sending it off to a journal, he submitted the report to Dr. Sellards for approval. But Dr. Sellards did not approve it. He believed his assistant's results were due to some error, possibly contamination, and he warned him that publishing the paper would be committing "scientific suicide." Friends advised Theiler that if he defied his professor by publishing, his chances for advancement at Harvard would surely perish.

Theiler had observed that when Dr. Sellards was good in the laboratory, he was very good, and his experiments were perfect. But when his ideas were bad, his experiments were also bad. It seemed to Theiler that Dr. Sellards's immense displeasure with him was connected, above all, to the success of his experiments—and the growing support they gave to his theory that a filterable virus, and not bacteria, was behind yellow fever.

Despite Sellards's negative response, Theiler submitted his report on the transformation of the wild yellow fever virus to the journal *Science*. The transformation was not something he professed to deserve much credit for; he had just tried something that had "happened to work." But that the results were solid he had no doubt.

Friends now advised Theiler that he did not have much of a future at Harvard—if he ever had. He was, they pointed out, thirty years old, and after eight years at Harvard, he was still only an instructor, at $3,000 a year.

Luckily, a better opportunity for him soon materialized. After his report was published in 1930, Dr. Wilbur Sawyer, the director of the Rockefeller Foundation's Yellow Fever Laboratory in New York, came to Boston to meet Theiler and learn more about his remarkable work. When he offered Theiler an invitation to join the Yellow Fever Laboratory at $6,000 a year, he departed from Boston with relief.

By 1928, yellow fever was once again an urgent concern. Since 1901 it had occasionally turned up in places where its presence could not be explained. There were reports now and then of yellow fever occurring in outlying areas where no *Aedes aegypti* were to be found. At first, little attention was paid to these reports of people contracting yellow fever in the middle of a forest or a jungle. But as the incidents continued, there was no mistaking them for anything but yellow fever.

People living near a jungle were sometimes victims; however, when a community or farmhouse was some distance from the jungle, only those who had actually been in the jungle caught the disease. Men, especially roadworkers and woodcutters, were stricken more often than women. It was reported that a child who had brought lunch to his father in the woods came down with it. Equally strange, people always seemed to believe that they caught it at midday. No sense could be made of these reports since *Aedes aegypti* mosquitoes are as domesticated as a Siamese cat and do not inhabit jungles, and the rare jungle species capable of spreading yellow fever under experimental conditions were never to be seen during the day.

There was another disturbing development. Just as *Aedes aegypti* mosquitoes seemed close to disappearing from the Western Hemisphere, due to diligent mosquito-control efforts, investigators came across *Aedes luteocephalus, Aedes apicoannulatus,* and *Eretmapodites chrysogaster*—three new species of mosquito, all capable of transmitting yellow fever. In 1929, *Aedes vittatus, Aedes africanus,* and *Aedes simpsoni* from Africa were added to the list. Then came *Aedes scapularis* from Brazil and *Aedes albopictus* from Java.

This was extremely ominous news: it seemed likely that there were many more players in the game than once supposed. And it certainly meant that yellow fever was not the simple problem it had once seemed.

One thing became clear with the mysterious appearance of jungle yellow fever: the plan to wipe out yellow fever by controlling *Aedes aegypti* mosquitoes had been a dream. If yellow fever lurking in the jungles were to be brought under control, it would have to be done with a vaccine. When Theiler arrived in New York, Dr. Sawyer's Yellow Fever Laboratory was hard at work trying to make one.

Most researchers at the Yellow Fever Laboratory did not notice Theiler or even know who he was. Five feet two inches tall, with a sallow complexion and dark circles under his eyes, Theiler kept mostly to himself, although people found him considerate, kind, and sensitive. He preferred to listen rather than to speak, a trait that tended to elicit long confidences, and when he disapproved, he did not usually say so. If conversation concerned an idea for an experiment, however, he invariably became animated.

When he appeared at the laboratory on Sixty-sixth Street and York Avenue in the morning, Theiler put on a stained lab coat and immediately went to check on his animals. Then he sat for several hours with his feet propped up on his desk, twirling a lock of hair and smoking, while he mulled over the day's plans and his technician made preparations for the afternoon's activities. After reading in the library and eating an early lunch, Theiler returned to the laboratory and worked for an hour or two. Then he relaxed until it was time to set out for Grand Central Station. One of his associates noted that he never seemed to expend any more energy than absolutely necessary to get his work done.

Dr. Sellards had demonstrated that while Theiler's tamed virus had lost its ability to destroy the viscera—the liver, kidneys, heart, and other organs—it had acquired on its travels through mouse brains a vicious affinity for the cells of the nervous system. Inoculated into the brains of monkeys, the tamed yellow fever virus produced fatal encephalitis. Thus, there was the danger that if it were given as a vaccine, it might infect the nervous system.

Dr. Sellards and a collaborator had actually immunized humans with the mouse-adapted virus, with some severe reactions. The method was later modified, producing immunity with less risk, but Theiler was horrified at this use of his virus. Dr. Sawyer also thought the method too risky to be used even on yellow fever workers, who desperately needed protection while they searched for a better vaccine. Nearly everyone who worked with yellow fever contracted it at one time or another, and the fortunate ones, who survived, were sick for weeks. Dr. Sawyer himself had been ill for more than a month with yellow fever.

Sawyer and two associates hit on the idea of injecting Theiler's tame viruses along with antibody-containing immune serum taken from recovered yellow fever patients, as protection against the virus's neurotropic tendencies. They ground up virus-infected mouse brains and combined the tissue with immune serum for the injection; as additional protection, a second injection of pure immune serum was given. In the spring of 1931, after testing it on monkeys, the vaccine was given to a doctor at the laboratory and other volunteers. It was safe, and it worked, and once it was administered to all yellow fever workers it ended laboratory infections.

This vaccine was not very practical because it required human immune serum, and, consequently, it became known as "the rich people's vaccine." Researchers at the Yellow Fever Laboratory agreed that it was necessary to further tame the virus so it could be used without serum.

Because the virus—the French strain Dr. Sellards brought from Dakar—had become neurotropic when Theiler grew it in mouse brains, Theiler and a German doctor, Haagen, at the laboratory decided to test a new technique that had just been developed by two researchers at Vanderbilt University: growing the viruses in chick embryos. After his hero, Adolf Hitler, came to power, Dr. Haagen returned to his homeland, and Dr. Sawyer replaced him with a young Canadian physician, Wray Lloyd, who had collaborated with Sawyer on the vaccine.

Theiler and Lloyd were opposites in disposition: Theiler was laconic and apparently unambitious, while Lloyd was a hard worker, expansive, and unabashedly ambitious. Their technician, Nelda Ricci, who was Italian born, rarely arrived at work on time, and anyone who attempted to reprove her did so at their own risk. Although her temper was volatile, Nelda Ricci was an expert at the delicate art of tissue culture; therefore, bacterial contamination, the bane of tissue-culture work in the days before antibiotics, rarely spoiled her experiments.

Under Nelda Ricci's direction, the viruses, tissue, and a nutritious "soup" were combined in sterile Erlenmeyer flasks, tightly stopped with corks wrapped in tinfoil, and kept warm at 37°C. for several days while, if all went well, the viruses multiplied. After centrifuging, some of the sedimented virus was transferred to new flasks with fresh medium to grow again.

The culturing of Dr. Sellards's French strain, begun by Theiler and Haagen, continued. Theiler and Lloyd decided that they would also try to culture the vicious Asibi strain. In 1927, Asibi, a twenty-eight-year-old black man, had been seen in a village about a hundred miles north of Accra on the Gold Coast. When Rockefeller Foundation doctors first encountered him, Asibi was resting with his head in his hands, feeling very ill. They diagnosed yellow fever and drew some blood. Returning to the village several days later, the doctors learned that their patient had recovered and gone back to work.

The virus they isolated from Asibi—the strain that had killed Adrian Stokes—was the most lethal yellow fever virus known. The Asibi strain was pantropic: deadly to both viscera and nerve tissue. Theiler and Lloyd chose it because they hoped that any change that might result from culturing would be easily detected. In addition, the Asibi strain had never been passed through mice, although it had been kept alive in the laboratory since 1927.

Theiler and Lloyd tried growing these two yellow fever viruses in every kind of tissue they could think of: whole chick embryo, minced chick embryo, small squares of embryonic skin, chick-embryo brains, tissue from adult mice and guinea pigs, tissue from embryonic mice and guinea pigs, even rodents' testicular tissue. The French strain multiplied readily, but the Asibi strain refused to grow until they offered it finely minced mouse embryo.

After culturing the Asibi strain eighteen times, they again tried growing it in minced chick embryo and chick-embryo brain, and this time, astonishingly, the Asibi virus multiplied. After fifty-eight subcultures, they tried chick embryo with the brain and spinal cord cut out on the theory that the virus might lose its penchant for nerve tissue if it were bred without it—as the tapeworm lost its mouth once it took up residence in the intestines and had no need for one. This notion was erroneous, but fortunately they did not realize it.

From time to time Lloyd or Theiler injected cultured viruses into mice or monkeys, looking for signs of change. They became discouraged when after a year the French strain was no more or less neurotropic than before.

But early in 1935 luck came from a most unlikely source: the virus in flask 17E—Asibi, grown in mouse-embryo tissue, which had been subcultured some ninety times. They injected it into the bellies of monkeys, and the monkeys *survived!* Its viscerotropism had utterly vanished! Equally remarkable to Lloyd and Theiler was the fact that the virus had also become less neurotropic than the modified French strain, on which they had pinned their hopes. They determined that the new Asibi virus was not sufficiently tame to be used without immune serum for vaccination, but it was the most well-behaved yellow fever virus anyone had ever seen.

Wray Lloyd prepared a report on their findings and included Nelda Ricci's name with his and Theiler's as the authors. Her skill had been important to their success, and she had done most of the actual tissue-culture work. She had carried on for weeks at a time while Theiler was busy with other projects and Lloyd was away, courting a Uruguayan woman, Chy-Chy Gonzales, whom he hoped to marry.

Dr. Sawyer did not approve of Nelda Ricci's name, or that of any other "nonprofessional" coauthor, on a scientific paper. Lloyd retaliated by tossing the manuscript into a drawer, where it remained for weeks, until Dr. Sawyer, eager to publish such important findings, relented.

In the fall of 1935, Wray Lloyd departed for Brazil to test the 17E vaccine, and Dr. Sawyer assigned Hugh Smith, a physician from South Carolina who had trained at Johns Hopkins, to replace him. That left Theiler, Hugh Smith, Nelda Ricci, and Old William—Vladimir Glasounov, a Russian who had fled his homeland in 1917. Before coming to New York, Glasounov had worked at the Rockefeller yellow fever station in West Africa, where he had converted to Islam because he had discovered that he could buy groceries at a better price that way. It was Old William who saw to it that the laboratory was always well stocked with mice, monkeys, and sterile glassware—equipment other laboratories found in short supply.

The work Lloyd and Theiler had begun with the Asibi strain went on. "It was a deadly chore," Theiler complained, but they seemed to be stuck with it. Before he left, Wray Lloyd had selected several strains of the Asibi virus he thought promising, and he had urged that more work be done with them. So Theiler and Smith continued to grow them in chick-embryo tissue with varying amounts of nerve tissue. They cultured Asibi with whole chick embryo (17D WC), Asibi with chick embryo brain (17D CEB), and minced chick embryo with the brain and spinal cord removed (17D).

No progress was being made with the French neurotropic strain—it remained exactly as it had been ever since Theiler had first passed it through mice in Dr. Sellards's laboratory—so they threw it out and began culturing the unmodified French strain and milder strains from Brazil.

"There was no logic here," Theiler explained later. "We set up a machine and let it run." Old William assembled the syringes, pipettes, and flasks for Nelda Ricci, who extracted the tissues from interminable thousands of mouse and chick embryos and prepared the culture media. Each time a new culture was started, a sample of the virus from the previous culture flask was removed, dried, and frozen. This was done by a third technician, who also inoculated mice intracerebrally at each tissue-culture passage to determine whether the virus was alive and well and still virulent.

Twice a week, Theiler or Smith injected specimens from each culture flask into mice, and they occasionally inoculated monkeys to test for change. As time passed and nothing happened, they grew bored and tested less often.

They kept themselves busy by trying to develop a ready supply of immune serum from Swiss goats for Dr. Sawyer's vaccine. Hugh Smith volunteered to receive the first injection of goat-serum vaccine; it made him sick, and after similar reactions in others, the goat serum was abandoned. The researchers then tried monkey serum, but it also presented difficulties. The culturing, tended mainly by the technicians, dragged on for a year.

On a spring morning in 1936, Hugh Smith arrived in the laboratory, glanced at the records, and noted that some of the mice inoculated with the cultured viruses had survived longer than usual before succumbing. He made it a point to recheck these experiments soon. When he did, he noted strange goings-on. Although the hind limbs of some of the inoculated mice had become paralyzed, the mice had not died.

A few weeks later, these mice were still alive! Smith also noticed mice inoculated with virus that were surviving without *any* signs of illness, and he mentioned this to Theiler. During the following weeks, they inoculated more mice and watched them closely. These mice also continued to dash about their cages as if nothing had happened to them. All the mice had been given the same virus—17D—one of the strains Wray Lloyd had selected before leaving for Brazil. It had been cultured first by Lloyd, Theiler, and Ricci, and then by Theiler, Smith, and Ricci—altogether it had been cultured 176 times.

Although Smith and Theiler remained cynical about the results, they decided an acid test for this virus was in order: monkey studies.

No one in the laboratory expected what happened next. Eight rhesus monkeys were injected subcutaneously with the 17D virus. All of them soon had mild yellow fever, but all eight survived—and within a week seven developed immunity!

When the 17D virus was injected directly into the brains of other monkeys, they did not catch encephalitis; although they became feverish, they quickly recovered. When 17D was injected into their bellies, they showed no signs of sickness. When wild yellow fever viruses were then administered to these monkeys, they still did not become ill. They, too, were solidly immune!

This, Theiler thought, was much too good to be true, and he repeated the monkey tests. The better the results looked, the more suspicious Theiler and Smith became. The history of vaccine-making, they well knew, was crowded with apparent successes that had faded ignominiously away, leaving a trail of deaths and ruined careers.

Hundreds of tedious tests of the 17D virus were needed, so Theiler was glad to escape the laboratory for a scientific meeting in London in the summer of 1936. Upon his return, he was not eager to go back to work. "I suppose the reason is that at present the work doesn't offer anything particularly exciting," he wrote in a letter to his mother. "What has to be done is so perfectly obvious that a good deal of interest is lacking. However, I have some very good assistants, so that I think we can manage a great amount of work in a short time and get the 'obvious work' out of the way and commence something new."

There had been no negative developments while he was away; so Theiler began trying hard to uncover some. He found that "however severe a test we give the vaccine, it comes through with flying colors." By November, Theiler and Smith decided the time had come to try 17D on themselves. Both were already immune to yellow fever, so there was not much of a risk; they primarily wanted to test for a reaction to the chick-embryo tissue. Following the injections, the researchers' antibodies increased; otherwise they had no reaction. Two nonimmune volunteers were then given injections, also with no ill effects, and both became immune.

As the test results accumulated, it began to dawn on Theiler and Smith that they *had* it! And they realized that it had been right under their noses for quite some time! Examining the records, they saw that at the 89th subculture, 17D had still been vicious, killing all three monkeys who had received it. No further monkey tests had been done until recently, so they took out the next viral sample that had been saved, the 114th subculture, and injected it into four monkeys. All survived without any signs of illness. Their calculations showed that the amazing mutation had taken place almost a year before they had noticed it! "The fact of the matter," Theiler said, "is that we were more interested in other things at the time."

Unfortunately, Wray Lloyd, who had done so much to bring it about did not share in the victory. He had died the previous summer in Brazil. Married to Chy-Chy Gonzales, the daughter of a Uruguayan physician, he was living in Rio de Janeiro when he fell from a window. According to one account, he had been hanging draperies in preparation for a visit with his in-laws; according to another, he had been sleepwalking.

The 17D viruses were tested by Hugh Smith in Brazil, where a vaccine was urgently needed, leaving Theiler to run the laboratory for a time with little assistance. The weekly quota of 1,500 mice continued to be delivered, and because he could not stop the deliveries and had nowhere to store the mice, he had to try to use them. He hardly had time to reflect that ten years before he had been forced to beg for six mice, and now he was about to be buried alive in mice.

In February 1937, when the eminences of the Rockefeller Foundation met to consider the new vaccine, they found no reason to believe it would not be absolutely safe for humans, so Theiler suggested they try it themselves. They did, with perfect results.

Theiler and Smith had just become convinced that 17D was not merely good but extraordinary when fate dealt Theiler a cruel blow. In March 1937 his eight-year-old son, Arnold, was hit by a car on his way home from school. The boy never recovered consciousness and died the following morning.

"And so ends the most happy little life," Theiler wrote to his mother several weeks later. "It's been awful. . . . People are very nice to us but their very kindness was at first the occasion of many exquisite and painful feelings." The arrival of flowers was most excruciating, reminding him and his wife anew each time of their loss.

Theiler decided not to take any time off from work, and he found a job in the laboratory for his wife, Lillian. Although they almost never spoke of it afterward, their son's death left them brokenhearted. His good job, their pleasant home—all seemed to lose meaning. The only reason for existence, Theiler finally resolved, was to do something worthwhile, "to leave the world a slightly better place to live in." He worked harder than ever and smoked even more than ever, emptying three packages of cigarettes a day.

By mid-March a dozen people, Lillian included, had been vaccinated. The immediate reaction to the vaccine was very mild— either nothing at all or only a slight fever seven days after inoculation—and everyone who received it developed immunity after fourteen days.

The news was announced in June 1937 in the *Journal of Experimental Medicine*. In their report, Theiler and Smith stated that the virus had lost its viscerotropism in mouse-embryo culture and its neurotropism through culturing in chick-embryo tissue with the nerve tissue removed. This, they explained, confirmed their theory—the original hypothesis on which the experiments had been based—that the amount of nerve tissue present had produced the remarkable mutation. "It would seem reasonable to conclude," they wrote, "that the changes in the pathogenicity and the tissue affinity that occurred during the prolonged cultivation are in all probability attributable to the different cellular components of the tissue-culture media."

Reasonable as this seemed, their conclusion was not correct. While Hugh Smith directed major vaccine trials in humans, Theiler repeated their laboratory experiments in their entirety to learn more about the mutation. This time he hovered vigilantly over the experiments, waiting for the change. But months, then years went by, and it never came!

Theiler was dumbfounded. Although the wild Asibi virus was cultured precisely as before, more than two hundred times during the next three years, it never again lost its deadly neurotropism. And Asibi virus grown *with* nerve tissue remained no more or less neurotropic than before. Other researchers also tried, but no one ever again produced a virus like 17D.

Theiler continued cultural promising strains in a great variety of media. But the more he learned, the less he understood. Although the French strain, cultured in the same media as before, soon lost its viscerotropism, its neurotropism actually *increased* in all three cultures, showing no correlation whatever to the amount of nerve tissue present. A mild strain from South

America decreased in neurotropism after 300 subcultures, but the least decrease occurred when the virus was grown with the least amount of nerve tissue. Other strains became so attenuated on repeated culturing that they failed to produce any immunity, even in large doses.

It was a humbling experience, and Theiler's only explanation of the strange, wonderful transformation of the Asibi strain, once the most deadly yellow fever virus of all, was: "There is no explanation."

The riddle of jungle yellow fever was solved one November day in 1940 by a falling tree. The search for the villain had led to the capture and study of tens of thousands of mosquitoes, and their every habit was scrutinized: from blood preferences to feeding patterns, flight ranges, sexual habits, life-spans, and preferred altitudes. There was a suspect: *Haemagogus spegazzinii*, a small, shiny blue mosquito that is a fierce biter. It inhabits jungles and coffee plantations, and the female drinks plant juices except when it is time to mature her eggs. Then she requires blood. The only drawback to the case was that *Haemagogus spegazzinii* is rare, and is never seen during the day.

On November 18, 1940, a team of researchers led by Dr. Jorge Boshell Manrique were in the mountains of Colombia studying yellow fever when they came across some woodcutters. The researchers stopped and watched as a large tree crashed to the ground. When the tree fell, its upper branches were shorn and it tore branches from surrounding trees as well, releasing clouds of *Haemagogus spegazzinii*, which promptly descended and attacked the workers.

At last the story was pieced together. In South America, *Haemagogus spegazzinii* inhabits the forest canopy and normally feeds on monkeys taking their treetop siestas. The monkeys are too alert at other times to be bitten. Partial to sunlight, these mosquitoes feed only during the day. The researchers realized

that yellow fever passes mainly between mosquitoes and monkeys, and probably between other animals and insects as well—
with humans of little importance to the survival of the virus.

In Africa a similar pattern was found. *Aedes africanus* feeds on
monkeys settling down for the night in the treetops; like *Haemagogus spegazzinii*, it seldom descends to the ground in quest of
human blood. But the monkeys descend in search of bananas on
nearby plantations, where they may be bitten by *Aedes simpsoni*,
which also bites humans, thus transmitting yellow fever.

After it was perfected, the 17D vaccine proved to be as good
as its makers first thought, and it put an end to epidemics of yellow fever. By 1960, 100 million people had safely received it, and
yellow fever was once again, as it had been before it traveled
west on slave ships, a rare disease.

In 1951, Max Theiler was awarded the Nobel Prize for his
work. Dr. Sawyer, who had directed the Yellow Fever Laboratory since 1935 and had developed the first good vaccine, did not
take the news well. Friends said he regarded it as a rejection of
his life's work. After the announcement of the award, Wilbur
Sawyer spoke of little else. He had already been ill, and at the
time he suddenly became much sicker. A month later he was
dead. His wife said that the award of the Nobel Prize to Max
Theiler, their neighbor and one-time friend, had killed her husband.

There were also those who said that Wray Lloyd and Hugh
Smith, as well as the Rockefeller Foundation, which had spent
millions to fight yellow fever, and many others, including William Gorgas, whose antimosquito campaigns first rid much of
the world of yellow fever, should have shared the prize.

For once Theiler was not reticent. He was as pleased about
winning, he said, as infielder Gil McDougald was when he hit a
grand-slam home run in the fifth game of the recent world series
and the Yankees beat the Giants, 13–1. He believed that Wray
Lloyd and Hugh Smith deserved much credit, but he didn't think

others should have shared in the prize. Other researchers had come and gone, he pointed out, while he was the only one who had worked continually in the field. And his mouse-modified French strain had made possible all the other vaccines.

When he was informed by reporters of the amount of money the prize would bring him, $32,357.62, he cracked a typical dry joke: "It looks as though yellow fever got me the *jack*pot." He told people who asked how he planned to celebrate that he was going to buy a case of Scotch and a season ticket to the Dodgers's games. But he never did. He continued to spend his leisure time reading, playing with his daughter Elizabeth, and watching baseball on television.

As doggedly as he had once studied yellow fever, he now worked on polio, and his studies of mouse poliomyelitis led to a new understanding of the disease in humans. This work was regarded by others as the most imaginative contribution of his career. Later, at Yale University, he became involved in a systematic, world-wide study of insect-carried virus diseases.

Theiler believed that the Nobel Prize changed his life in only one respect. At the train station in Hastings-on-Hudson, where he used to catch the New York Central, he was known by his neighbors as "the man who lives next door to Alvin Dark" of the New York Giants. Afterward, he became "the Nobel Prize winner who lives next door to Alvin Dark."

Yellow fever never did get Theiler; he died of a heart attack at the age of seventy-three in 1972.

3

The Sculptor

WHEN GERMAN TROOPS INVADED Bialystok, Russia, in August 1914, an eight-year-old Jewish boy fell into step and playfully marched along with them to the center of the town. There, on the courthouse steps, sitting as close as he dared to one of the enemy soldiers, the boy watched him pull out a small dark loaf and spread it thickly with butter. The city had been besieged for some time, and Albert Sabin thought he had never seen a sight as wonderful as that German pumpernickel. He would never forget that pumpernickel, or the soldier who shared it with him.

Sabin would remember, too, the great relief of Bialystok's large Jewish population over the German occupation. His parents noted how much more civilized the Germans seemed than the Cossacks, who had just departed, and when a German school was opened shortly thereafter in Bialystok, the boy was enrolled there.

Few Sabins remained in Bialystok at that time. After the pogrom of 1905, all Albert's surviving relatives had left the textile center for the United States, except for his immediate family who had stayed to look after his mother's ill parents. There were four children in the Sabin family, two younger sisters, an older brother, and Albert. Their father worked as a weaver, at home. They were not desperately poor—there was money for the German school—but life was not easy.

In 1920, when the last family ties were severed, the Sabins packed their belongings and set off for the United States. For months they made their way through Europe, from Warsaw to Danzig to Antwerp, where they were finally able to board a boat for New York. By then, they felt fortunate to travel in steerage.

The family settled with relatives in the silk weaving center of Paterson, New Jersey, and Albert's father and older brother found work in the silk factories. Albert remained in New York for six weeks to be drilled in English, American history, mathematics, and literature by two cousins who were successful civil-engineers. Albert knew some English when he arrived in the United States, and by February 1921 he felt prepared to go to school. He presented himself to the principal of Paterson High School with a diploma from the Russian school he had attended. But the principal couldn't read Russian.

"It means," said the fourteen-year-old Sabin, "that if my education had not been interrupted by emigration to the United States, I would now be a sophomore." The principal went along with this, but warned Sabin that he had better do well in his studies. He did do well, with coaching from his cousins, but he had to work after school and on Saturdays and studied late into the night. He had no time for sports, although he did join the debating society, which he relished.

Sabin graduated in 1923, at the age of sixteen, and he thought of studying law, but he had no money for college. Then his uncle, a busy dentist, offered to pay his college tuition and provide him with room and board in his New York City apartment if Albert would become a dentist and assist him during his years of training.

Shortly thereafter, Sabin was enrolled not only in New York

University's Washington Square College but also in a school for dental technicians, and again he worked long days and evenings. After a year at Washington Square College and two years at the New York University College of Dentistry, Sabin decided that he didn't like to fix teeth.

Bacteriology interested him far more, and reading Sinclair Lewis and Paul de Kruif, he was enchanted by the microbe hunters. He liked to imagine himself as one, and pushed aside the disquieting thought that all the great work in bacteriology had already been done—"What is there left for me to do? Gild the lily?" He made up his mind to a career in medical research.

Hearing of his plans, his uncle ordered him out, in eleven memorable words: "You don't work for me, I don't take care of you." So Sabin again worked at odd jobs to earn money and took premedical courses at Washington Square College. He then attended New York University College of Medicine. Although he had a scholarship, he worked after school at Harlem Hospital to earn his room and board.

Dozens of pneumonia patients were admitted to the hospital every day, and it was Sabin's job to inoculate mice with the patients' sputum. By the following day, the mice would be dead, and Sabin would identify the type of pneumococcus that had killed them so the proper antiserum could be administered to the patients. But by then many of the patients were also dead.

"My God," Sabin thought, "should a person have to wait a whole day for treatment when it is a question of life or death?" It occurred to him to try inserting a capillary tube into the mouse's belly at hourly intervals, after injecting the sputum, to see when enough pneumococci had appeared to determine their type. He found that the organism could be identified within three hours, which meant that treatment of the patient could begin a whole day earlier than before!

Many lives were saved by this procedure, and Sabin was elated. But his elation did not last long: he soon saw that even when treatment was begun earlier, some of the patients still succumbed for no apparent reason. He set about learning why, and by the time he graduated from medical school in 1931, he was immersed in the study of pneumococcal infection and had published several papers on his discoveries.

In July of that year a major polio epidemic swept New York City, paralyzing 6,500 persons during the summer and killing many of them. Dr. William H. Park, chief of the bacteriology department at New York University, where Sabin was working, suggested that Sabin investigate polio; but Sabin was preoccupied with pneumonia and protested that he didn't know much about viruses anyway, other than what he had read. Dr. Park had liked Sabin's clever method of identifying pneumococci and wanted him to look into a new skin test for polio immunity developed by Professor Claus Jungeblut of Columbia University. So he offered incentives to the ambitious young man: money, more laboratory space, and precious monkeys.

Sabin promptly looked into Professor Jungeblut's test and reported to Dr. Park that the professor was working with an "awful mess of stuff," and he didn't see how his test could prove anything. It was an impudent remark, but sure enough, when Sabin repeated the professor's test, the results could not be confirmed. So Dr. Park encouraged Albert to try purifying some poliovirus. Using the methods of the noted German chemist Willstätter, of Munich, who had purified enzymes, Sabin went to work.

Sabin succeeded, much to his surprise, since this had never been done before. Using the purified poliovirus, he again tried to repeat Professor Jungeblut's test and failed. He went to the professor and asked what he might have done wrong. They repeated the test together, and still it failed to work. Sabin published the bad news, dashing all hopes that the test would be a weapon against the dread disease.

Almost nothing was known about polio. There was no preventative, no cure, and no one knew why the disease was becoming epidemic. Poliolike diseases had been described throughout history, but no epidemics had been reported until late in the nineteenth century, in Sweden and the United States. The first great epidemic in the United States did not occur until 1916, when almost 30,000 people were stricken and thousands perished.

In its most devastating form, polio strikes the base of the brain, rapidly causing coma and death. If the victim survives, paralysis of the respiratory muscles sometimes develops, and these victims are confined to iron lungs, which draw their every breath. Since these victims cannot cough—and the iron lung cannot cough for them—pneumonia eventually sets in and they die. No one survives an iron lung; death comes in a year or perhaps in ten.

In less severe cases, the lower spinal cord is affected, with the virus attacking the anterior horn cells, which provide the nerves of the arms and legs, and victims are left with paralyzed, useless extremities. The disease often struck children, and the twisted, deformed bodies of these small survivors were especially saddening.

The alarming trend toward more frequent and widespread polio outbreaks was continuing, and no one, including the most brilliant of Sabin's professors, had any idea of what to do about it. Sabin savored publishing his paper on the challenging subject.

As an intern the following year at Bellevue Hospital, Sabin was distracted by other matters. A researcher he had known at the City Health Department Laboratories had died suddenly, after being bitten on the hand by a monkey. Sabin obtained some autopsy material and isolated a virus from it. After studying it, he announced that he had found a new virus. A professor at Columbia University disagreed, arguing that it was the well-known human herpes simplex virus.

But this virus, which was harmless to monkeys and fatal to humans, proved to be a new one after all, and Sabin called it herpes B virus after its victim, Dr. Brebner. "Maybe because I wanted it to be a new virus I said it was a new virus," he remarked later. "I was naive. If I had known more I would have said it was herpes simplex."

More important than being correct, although he did not mind that in the least, was the lesson that lodged in his head: that evolutionary forms of viruses exist in nature. The relationship between herpes simplex and herpes B virus was close enough to suggest that they had a common ancestor, and that through evolution they had diverged slightly. Yet both forms existed at once: one less virulent than the other.

Because of Sabin's work with the B virus, he was invited to join the Rockefeller Institute. Not yet thirty, he arrived at the institute in January of 1935, fresh from a year at London's Lister Institute, wearing a tweed jacket and smoking a pipe.

"God, he was a sight when he arrived," recalled Tom Rivers, who had found him the job. "He was the most elegant dresser in the entire institute." He also proved to be, as Rivers had assured the institute's director, "a nice young Jewish boy who is as smart as all outdoors." He knew a great deal about viruses, and Rivers found that Sabin "quickly showed that he was capable of doing elegant work in the laboratory."

Sabin, now married, did not have any children, but he became more interested in polio. It was a devilish challenge, and no real information had been gained since the institute's director, Dr. Simon Flexner, had transmitted the disease among monkeys in 1909.

The greatest obstacle was that the virus could not be grown in the laboratory. Sabin and his boss, Peter Olitsky, tried culturing the virus in different human tissues and discovered that it could indeed be grown in culture—but only in human nervous cells. Their experiments, which were done with the utmost care and precision, were hailed as a major breakthrough. Neither the researchers nor anyone else suspected that the Rockefeller Institute strain of poliovirus they had used in the studies had led them into a trap.

Another important concept about polio established at about the same time, in the mid-1930s, was that the virus invades the human body through the nose and from there travels its deadly course along the olfactory pathway to the brain. This was supported by Sabin's findings of incriminating lesions left by the virus on the neurons of monkeys' olfactory bulbs. Some researchers concluded, therefore, that a nasal spray of certain chemicals might prevent polio; and in monkeys this seemed to be the case. When children were tested, however, the nose spray failed to prevent the disease, and some of the children permanently lost their sense of smell.

Two far more calamitous efforts at preventing polio followed in 1935. Inspired by Pasteur's rabies vaccine, made fifty years before from the spinal cord of rabies-infected rabbits, two American researchers made separate attempts to produce a vaccine for polio. Maurice Brodie, at New York University, infected monkeys with poliovirus, killed the monkeys and ground up their infected spinal cords. This substance he treated with formalin to inactivate the virus, before injecting it into monkeys. Dr. Kolmer, of Philadelphia, also worked with monkeys' spinal cords, but they contained live, untreated viruses he believed he had weakened—as Pasteur had weakened the rabies virus—by transmitting them repeatedly from animal to animal. Brodie and Kolmer tested their vaccines on a few monkeys before administering them to thousands of children. In both instances, a few vaccinated children were stricken with polio and died.

It was a tragedy not only for the victims and the scientists—Brodie later killed himself—but also for the cause, for it demonstrated the extreme complexity of the enemy. "The correct question," Sabin saw, "was not *how* to make a polio vaccine, but rather, *could* a vaccine be made?"

Sabin believed that polio called for a full-scale military campaign. He had grown up in the midst of war (after the Cossacks and Germans left Bialystok, the Polish army arrived, followed by the Bolsheviks, all before the region was restored to Poland in 1919) and he was well aware that the first step in a successful campaign is reconnaissance. "It would be necessary to determine

the enemy's strengths and weaknesses, its hiding places, its different faces. How many different kinds of polioviruses were there?
Where in nature did they multiply in the numbers necessary to
ensure their continued existence? Where in the human body did
they multiply? How did they invade and escape from the body?
How many different kinds of immunity did they produce? How
long did immunity last?"

The failure of nose sprays to halt polio had made Sabin doubt
the olfactory-pathway concept. Perhaps, he thought, polio in
monkeys was one thing and polio in humans was quite another.
"Human beings would have to be studied," he decided, "to understand the human disease."

So Sabin did something quite simple: he examined the olfactory bulbs of a polio victim. Not finding any lesions, he examined more cases, and still there were no signs of lesions. After
further studies he was forced to conclude that in human beings
polioviruses do not enter through the nose or multiply mainly in
the nose, as was widely believed, but rather they first multiply in
the intestines.

How did they wind up there? By way of the mouth most
likely, he thought; but it was ten years before this theory was
confirmed. Sabin and researchers at Yale and Johns Hopkins universities fed freshly isolated strains of poliovirus to monkeys and
chimpanzees and painstakingly charted their progress. From the
mouth, they learned, polioviruses progress to the intestinal tract,
where they settle in the intestinal lining to multiply. There, the
viruses are routed by the lymph system into the bloodstream,
where they multiply in other non-nervous tissues before invading
the spinal cord and other parts of the brain. Finally, and here the
stunning insidiousness of the scheme was at last revealed, some of
the live viruses are excreted and set free, ensuring that all the
viruses will not be buried along with their victims.

Sabin began to wonder if the idea was not to try thwarting
such an awesome natural scheme, but to use nature to somehow

get around nature. He began to wonder if vaccine viruses might follow the same route as the wild viruses. . . .

As the years passed, polio epidemics worsened. Tens of thousands died of the disease, and by the late 1940s there were hundreds of thousands of crippled survivors, many of them children. Some were destined to spend the rest of their lives in iron lungs; the others could be seen on the streets of almost every city and town in the United States. The disease was not as great a killer as influenza, but these survivors were a constant reminder of the disease. Partly because of them and partly because of the National Foundation of Infantile Paralysis, the battle against polio became the greatest voluntary crusade against a disease of all time.

The National Foundation of Infantile Paralysis, founded in 1937 by President Franklin Roosevelt, himself crippled by polio, was directed by the president's friend, Basil O'Connor, a Wall Street lawyer. Under O'Connor's flamboyant supervision, the foundation's appeals, notably the March of Dimes, raised billions of dollars—about $25 million a year—between 1938 and 1972. "My name is virus poliomyelitis," began one of the foundation's famous fund-raising films. "I consider myself quite an artist, a sort of sculptor. I specialize in grotesques . . ."

Among the many researchers who were supported by foundation funds, including Albert Sabin, Basil O'Connor favored a young professor at the University of Pittsburgh, who offered the most immediate prospect of producing a vaccine. He was Jonas Salk. Like Albert Sabin, he was young, energetic, smart, and aggressive, not one to waste time when it came to pursuing a promising opportunity.

In 1949 the result of a vast study comparing viral samples from thousands of polio victims was announced: although many strains of poliovirus had been found, there were only three different antigenic types. This fact—that there were only three kinds of virus to immunize against—meant that making a vaccine suddenly seemed much more possible.

That same year Sabin and Olitsky's classic studies showing that poliovirus could grow only in nervous tissue were overturned by Dr. John Enders and two associates at Children's Hospital in Boston, who worked with different strains of the virus. They used foreskin tissue for their first experiments and later grew the viruses in monkey kidneys. This discovery, for which they were awarded a Nobel Prize, was a critical breakthrough, since viruses prepared for vaccine in nervous tissue may induce neurologic disease—a lesson painfully learned by previous vaccine-makers.

Sabin and Olitsky had experimented with a strain that had been passed through monkeys' brains so often it simply lost its ability to multiply in non-nervous tissue. "Work with monkeys led to many misunderstandings," Sabin would say ruefully long afterward, though he could chuckle over "the luck that I happened to work at the Rockefeller Institute and got my hands on Simon Flexner's old poliovirus that had been around since *1909!*"

If Sabin and Olitsky had worked with a different strain of poliovirus, virologist Tom Rivers felt, "the chances are that they would have been able to grow the virus in non-nervous tissue, and we would have had a breakthrough of major proportions in making a vaccine. As it was, we had to wait fourteen years for this particular breakthrough."

Almost everything that was learned about polio in the next several years continued to point to the possibility of making a vaccine. A study of isolated Eskimos revealed that antibodies to polio could last very long, for forty years or more. Dr. Dorothy Horstmann of Yale University found that in humans poliovirus enters the blood during the early stages of the disease, suggesting that small amounts of circulating antibody might be able to destroy the virus before it reached the central nervous system.

(It had been thought that the virus did *not* enter the blood, because it had left the blood and invaded the brain by the time anyone got around to looking for it.) It was also learned that a poliovirus infection not only causes antibody formation in the bloodstream but also resistance to infection in the intestines—pointing to the possibility of eliminating the viruses' very breeding ground—the intestine—with a vaccine before it ever reached the bloodstream.

Most compelling, Albert Sabin thought, was the fact that several strains of poliovirus had been tamed. By the early 1950s, Max Theiler, who tamed the yellow fever virus, had modified a strain of poliovirus by transferring it from monkeys to rats and then to mice. When fed again to monkeys, this virus did not cause paralysis. Dr. Enders and other researchers had also produced experimentally weakened strains, and this greatly impressed Sabin.

"A poliovirus is not a poliovirus," he kept repeating to himself, and he gradually determined that the correct approach would be to take "hot" viruses and tame them or to select from nature "good" viruses, i.e., viruses that multiplied in the intestinal tract and stimulated the production of antibodies but that multiplied very little, or not at all, in the human nervous system. These good viruses, which would still resemble their vicious relatives sufficiently to stimulate antibody formation in the blood and resistance in the intestines, would be *fed* to people in imitation of a natural infection.

Hunting down or breeding such viruses would be an enormous task. He would be looking for not just one good virus, as Max Theiler had been, but one for each of the three different antigenic types that caused polio. But Sabin saw no good reason it shouldn't work.

"Obviously, factors involved in human evolution played a role in the evolution of polioviruses," he reasoned. "So, if by natural selection polioviruses with certain traits advantageous to the poliovirus came to dominate—and they did, clearly—then by experimental selection, why not try to select polioviruses which would favor humans?"

No one had ever immunized people by feeding them live, infectious viruses, but the histories of smallpox, rabies, and yellow fever showed Sabin that only live viruses made a good vaccine.

It was an audacious idea, but then he was frequently accused of being an audacious person. Before Sabin arrived at the Rockefeller Institute, it had been rumored that he would appropriate all the monkeys in the place for his own use, and colleagues later complained that Sabin demanded and usually got things they would never think of even asking for. "A supreme egotist," commented one associate. "A very unpleasant individual," said another. But a third attributed Sabin's unpopularity to his habit of saying exactly what was on his mind: "He could see much further than [his critics] and could see whatever he looked at more thoroughly and clearly, and thus could promptly relate the significance of a problem to the world at large. In his forthright, confident manner he would explain to them, and . . . they would resent being given the answer by a tyro much younger than they."

Jonas Salk had set promptly to work, in 1950, experimenting with the many new techniques for culturing viruses in monkey tissue. Before long he was successfully harvesting large quantities of virus, and he announced that he could produce a safe, effective polio vaccine. Like the influenza vaccine he had worked on during the Second World War, this would be a "killed"-virus vaccine. He spent the early 1950s searching for suitable strains of each of the three types of poliovirus—strains that were virulent enough to promote good antibody formation, but ones that were easily inactivated. The viruses would also have to grow well in culture. For Type III poliovirus, Salk selected a strain that had been isolated from British soldiers of the Middle East Forces, and for Type II he chose a virus isolated from a young American polio victim, James Saukett.

For Type I, Salk settled on the notorious Mahoney strain, as deadly a poliovirus as anyone had ever encountered. It had spread from a family named Mahoney in Akron, Ohio, to some of their neighbors, whom it had killed. Salk was criticized for including the Mahoney strain in a vaccine, but he liked it; the Mahoney strain, once inactivated with formalin, still stimulated antibodies, and it thrived in tissue culture.

Next, Salk labored over the process that would inactivate the viruses, to be certain that not a *single* viral particle remained alive. Unlike Maurice Brodie's disastrous methods, Salk's procedure was highly scientific. He carefully studied the amount of formalin necessary and the time required to "cook" the viruses and developed an inactivation curve, to determine the precise point at which all the viruses would be dead. The trouble was that as the process continued and the number of active viruses decreased, they became ever harder to detect, so he added more time to the procedure as a safety margin.

Sabin and others strongly disapproved not only of Salk's use of the extremely dangerous Mahoney strain in a vaccine but also of the inactivation procedure. But Salk remained adamant on both points.

Sabin, now at the University of Cincinnati College of Medicine, was busy searching for docile viruses. Experimenting with one strain after another, he grew the viruses in monkey-kidney cultures, harvested them, and cultured them again and again, altering conditions to favor the appearance and overgrowth of viruses with very precise characteristics: nonvirulent but antigenic and able to proliferate in human intestines.

Sabin postulated that when advantage was given to viruses that multiplied first in non-nervous cells, he would obtain viruses that were less virulent. But he found that some of the progeny of these viruses were tame and others virulent. So he began diluting out the undesirable viruses, theoretically to the point at which they would be completely gone. The remaining viruses were cultured and inoculated into the brains of monkeys to test for virulence. It was slow work, tampering with evolution.

Jonas Salk's vaccine progressed much more rapidly. Early in 1953, after he had tested his killed viruses on monkeys, Salk informed the National Foundation of Infantile Paralysis that he had also tested his vaccine on small numbers of institutionalized children in Pennsylvania—with excellent results. Critics thought Salk's vaccine was far from ready for large-scale testing, but the National Foundation disagreed, and preparations were begun for nationwide field trials involving hundreds of thousands of children.

Later in the year Sabin informed the foundation of his own good news: he had nonvirulent forms of all three types of poliovirus in hand. When he fed the new viruses to monkeys and then inoculated the monkeys' brains with lethal doses of wild poliovirus, they did not become ill. He suggested to the foundation that this approach might produce a better vaccine than Jonas Salk's.

Sabin was informed that the foundation would support Salk's vaccine. Thoroughly annoyed, Sabin went back to work.

In the spring of 1954, as nationwide trials of Salk's vaccine were about to begin, Sabin learned something very important. When he inoculated the gray matter of monkeys' spinal cords with his new mutant viruses, paralysis occurred, but when he inoculated chimpanzees in the spine with the same viruses, they did not become ill. The nervous system of monkeys, which are lower primates, in other words, was more susceptible to poliovirus than the nervous system of chimpanzees and by analogy, humans!

Yet the reverse was true for the intestinal tract; the human intestinal tract was easily infected by doses of poliovirus that were ineffective in chimpanzees, and the chimpanzee's intestinal tract was in turn more susceptible than that of monkeys. Further, the two traits, nonvirulence for the nervous system and the ability to multiply in the human intestines, were not genetically linked to each other! What more could one ask of nature!

With this discovery, Sabin knew he had it. He had isolated polioviruses that were potentially safe for humans, and he now knew how to carry out the necessary tests to prove it.

The National Foundation of Infantile Paralysis did not concur. The foundation's virologists pointed out that no one knew how stable these new viruses were and expressed concern that once they were unleashed from the laboratory, they might revert to their old ways.

Sabin was enraged. Tom Rivers, head of the foundation's vaccine committee, recalled in his memoir: "He just about went up like a skyrocket."

A few weeks later, Sabin recovered himself enough to demand a thousand monkeys. "What do you do, eat the monkeys?" Rivers shouted, and he again said no. Sabin had already used a thousand monkeys that spring, Rivers reminded him, and monkeys, which were scarce to begin with, were needed now to produce Salk's vaccine.

Sabin remarked that the foundation had supported him only as long as they had thought his vaccine wouldn't work. He wanted to tell Rivers and Basil O'Connor and his foundation exactly what he thought of them, but he knew that working on polio in the United States without their sanction was unthinkable.

Again in the fall, Sabin asked the foundation for permission to administer his viruses to human subjects, and after more infuriating months of delay (while testing of Salk's vaccine proceeded) permission was finally granted.

When he administered his three strains to susceptible prisoners in an Ohio penitentiary during the winter of 1954, no one became ill, and all developed antibodies in their blood and resistance in their intestinal tract.

Then on April 12, 1955—the anniversary of President Roosevelt's death—the outcome of the Salk vaccine trials was announced by the National Foundation of Infantile Paralysis at the University of Michigan. There were 150 television, radio, and newspaper reporters in attendance, and 16 television and newsreel cameras recorded the proceedings. The occasion, the *New York Times* noted, was more in keeping with a Hollywood premiere than a great moment in medical history.

It was reported that 440,000 children in forty-four states had received the vaccine and that it had been found safe and effective. The vaccine was, in fact, being licensed that day by the Public Health Service. Huge supplies of the vaccine, already on hand in anticipation of the event, were distributed across the country. Word spread around the world that polio had been defeated. Jonas Salk was a hero.

Two weeks later, after almost half a million children had been inoculated, there was shocking news: a few of the vaccinated children were ill with polio. No one knew how many might be affected or what had gone wrong. The program was quickly halted.

Several hundred vaccine-related cases of polio resulted; 150 children were paralyzed and 11 children died. Federal investigations were held, and accusations flew.

The immediate problem was determined to be Salk's inactivation procedure, which when practiced on a smaller scale for the field trials, had killed the viruses. But when formalin was added to large batches of virus, precipitated protein had formed a shell around some of the viruses, protecting them from the action of the chemical.

The questions of how the live viruses had eluded safety testing and whether the Mahoney strain in particular was at fault were unanswered and became the subjects of lawsuits that were not resolved for decades. The firm that produced the faulty vaccine, Cutter Laboratories, was accused of carelessness. Jonas Salk and Basil O'Connor were blamed for their haste in giving the public a vaccine and for the secrecy that had surrounded the research.

After more stringent safety-testing procedures were instituted, vaccinations were resumed. But Albert Sabin, John Enders, and other virologists argued that the program should not resume until less virulent viruses could be tested.

No matter how often he diluted his new viruses, Sabin could not manage to purify them completely. Viruses that were sufficiently tame did not always multiply well, and those that did multiply sometimes partially reverted to virulence in later generations.

It did not help Sabin that many people seemed hell-bent on convincing him that his vaccine was a failure. After the tests on prisoners, when he learned that some of the viruses were reverting, Tom Rivers, his old friend, said, "Look, why don't you throw this stuff down the sewer and forget it?"

At about that time a new technique was introduced by the eminent virologist, Renato Dulbecco, for obtaining the progeny of single virus particles, and Sabin saw a potential use for it. Rivers organized a conference early in 1956 to discuss the technique.

"Usually when Sabin attends a meeting or conference," Rivers said, "he does the telling, and more often than not it is worthwhile because Sabin is a very smart hombre." But this time he invited a lot of "heavy guns"—Renato Dulbecco, Joshua Lederberg, MacFarlane Burnet, Max Delbrück, and Salvador Luria—to do the talking.

Sabin did listen, and following the conference he took his three best strains and went to work with Dulbecco. In petri dishes they placed individual viral particles on a thin layer of cells and fixed them in place with an agar gel. Each virus produces a colony (clone), which is called a placque because it is essentially a hole in the cell layer. Thus each placque contained the progeny of a single parent; and Sabin found, by testing them on large numbers of monkeys, that his strains were indeed not homogeneous.

With repeated placquing he had what he wanted before the

end of the year: three strains that were less virulent than any he had previously tested. When he inoculated these viruses into the ultrasensitive gray matter of monkeys' spinal cords, they only rarely caused mild paralysis, and when very large doses were given, the paralysis did not spread. Even the foundation approved of his viruses, and he resumed testing them on volunteers.

By 1957 he had carried out tests on 10,000 monkeys, over a hundred chimpanzees, and several hundred prisoners, and Sabin had even eaten the viruses himself—there was little danger in this since he already possessed some immunity to polio. But so far all the subjects had been adults, and because the vaccine would eventually have to be given to children, it would first have to be tested on unimmunized children.

Sabin thought of his daughters, Debbe and Amy, who were five and seven years old. He knew they had absolutely no immunity to polio because he had tested them himself—and he wondered if he could give them the vaccine.

Although he was invariably much too busy to be a good father or husband, after fourteen childless years of marriage, he did not take his children lightly. He had been so pleased at becoming a father, in fact, that he imprudently confided to reporters later, at the Hotel d'Angleterre in Copenhagen, that his first child had been conceived in that very hotel. Although he had been speaking strictly off the record, the story about his "Danish baby" appeared on the front page of the Copenhagen newspapers the next day, while some news about his vaccine was barely visible in a lower corner.

He knew he had reached the turning point: if he were not prepared to give the vaccine to his own children, he could never give it to anyone else's. It would be the end of the road for his vaccine. He and his wife, Sylvia, discussed it repeatedly, and he spent many sleepless nights reviewing the rationale for what he was thinking of doing. He had worked for twenty-five years to reach this point, taken every conceivable precaution, tested the strains in every possible way. But a slight risk remained. All risk could never be eliminated.

To make the matter even more difficult, others would be involved. Sylvia, who had been raised on an Illinois farm, had no immunity to any of the three types of polio, nor did their neighbors or their neighbors' three children, who would also be put at risk.

Finally, his wife and the neighbors were willing, and he couldn't find a good excuse for not going ahead. Sabin fed the first type of weakened poliovirus in cherry syrup to his daughters, wife, the neighbors and their children. His daughters cried as he swabbed their throats, stuck needles in their arms, and demanded specimens of their excretions, and it was many years before they forgave him for this ordeal. He did this three times a week for many weeks; then he repeated the procedure all over again after giving them the second and third types.

Debbe, Amy, Sylvia, and neighbors all remained healthy, and the viruses behaved in them exactly as they had been bred to, proliferating in the intestines and triggering antibody formation and intestinal resistance. Viruses retrieved from their feces seemed to be highly attenuated.

After additional tests were done on children in the United States, Mexico, Europe, and the Soviet Union, the next logical step was a large-scale field trial. But Sabin faced the immediate problem of where to hold it. There were few communities in the United States where none of the children had received the Salk vaccine. As Sabin was considering this, he received distressing news from the foundation: they were not going to approve a field trial of his vaccine in the United States anyway. There were still concerns about reversion of the viruses to virulence. A year before, it was pointed out to Sabin, Dr. Hilary Koprowski of Lederle Laboratories had tested attenuated polioviruses in Ireland, and two strains had dangerously reverted to high neurovirulence.

Sabin was furious. His strains were different from Koprowski's, and he had demonstrated that his strains would *not* revert significantly. He realized that if the National Foundation prevailed, his vaccine might never have a chance. But he did not know what to do.

Sabin was at wits' end when, in 1957, the World Health Organization assumed authority over all polio immunization and promptly endorsed field trials of his vaccine. But the problem of where to hold them remained. Dr. Koprowski had tested his vaccine in the Congo and Ireland, and other attenuated strains had been tried in Central and South America. Most other developed countries had administered the Salk vaccine.

Then Sabin recalled how smug Russian scientists had been at international conferences a few years back over the absence of polio epidemics in the Soviet Union. He had been amused when a Soviet delegation subsequently arrived in the United States to learn how to make the Salk vaccine. Sabin invited the Russians to visit his laboratory in Cincinnati. When they did, he described to them the many advantages of his live vaccine and offered them the opportunity to try it in their country. He was soon invited to the Soviet Union, and arrangements were made for a collaboration.

Sabin was impressed by the speed with which the Russians got to work. They conducted trials on hundreds, then thousands, then millions of children. He knew that his vaccine was good, but he did not know until the Soviet study was completed how practical it was on a large scale.

In June 1959 at a World Health Organization conference in Washington, D.C., Sabin summarized his own studies and reported the results of studies done on 4.5 million people who had received either a vaccine from Sabin or from secondary lots prepared in the Poliomyelitis Institute in Moscow. Everyone vaccinated showed an immune response and none became ill with polio. In vast, meticulous follow-up studies, the Soviets reported no ill effects either on the unvaccinated in any of the communities where immunization had taken place.

The report on Sabin's vaccine was so impressive that it aroused suspicion. The suggestion was made by some government officials and scientists that the Russians might not be as careful about their research as the Americans. So teams of American scientists were dispatched to the Soviet Union in 1959 and 1960 to assess the trials; but they uncovered no flaws.

Testing continued, particularly because of the question about reversion to virulence. Sabin found that his modified viruses, after passing through volunteers more than half a dozen times, showed a slight increase in virulence when he injected them into the ultrasensitive spinal cords of monkeys. But feeding the virus by mouth is very different from injecting it directly into the nervous system, and this was not judged to be a serious concern.

Still, more information was needed, and Sabin personally directed a community-wide test of the vaccine in Cincinnati in 1960. About 200,000 children, their families, and the population of the surrounding region were studied. Some of the children received vaccine, while others did not. Then the effects of the spread of the vaccine viruses within the homes and the community were observed. With the cooperation of police and health officials, anyone who developed any kind of neurologic symptoms during the period of the study was brought to Sabin's attention and thoroughly assessed. "These were heroic studies," Sabin himself did not hesitate to point out. "Nothing like it had ever been done before."

Already, more than 6 million people worldwide had been successfully immunized with his vaccine, and in December of 1959 the Ministry of Health of the Soviet Union had approved the vaccine for general use. In 1960, millions in the Soviet Union and the satellite countries received the three types of live vaccine.

In the United States and other countries Sabin's vaccine had dramatically halted epidemics, and Sabin was becoming very impatient. His vaccine had been studied more thoroughly than any other in history—Salk's vaccine had been tested on 400,000 people before being licensed—and his oral vaccine was already widely accepted and used abroad, often to the exclusion of Salk vaccine. Yet in the United States, only the Salk vaccine was licensed for use.

Sabin began lobbying loudly and incessantly for his vaccine, pointing out its many proven advantages. The live vaccine gave two kinds of protection: it stimulated the formation of antibodies in the bloodstream, like killed vaccine; but it also produced intestinal resistance to infection, while killed vaccine produced little if any intestinal resistance. Thus, only live, oral vaccine would reduce the chance of intestinal infection by wild polioviruses and would eliminate the very real danger of vaccinated persons' transmitting virulent poliovirus, from their intestines, to the unvaccinated.

Furthermore, live vaccine caused antibodies to form much more quickly than killed vaccine—in days rather than in weeks or months—which meant that live vaccine was useful during an epidemic whereas killed vaccine was not. Live vaccine was also less expensive to make and administer and induced longer-lasting immunity than killed vaccine. Thus, it would be more useful in poorer countries and stood a better chance of being effective.

Further, the live vaccine conveyed immunity to some who had never received it by passage of the attenuated viruses from vaccinated to unvaccinated. Thus, if enough people were to receive it, live vaccine might alter the very ecology—with the attenuated viruses multiplying to the extent that they would outnumber the wild viruses. "With live vaccine, you don't get rid of polioviruses," Sabin said. "You get rid of the viruses that cause paralysis!"

It was an extraordinary concept, Sabin pointed out—there was no other vaccine like it, and he failed to understand why there was such reluctance to accept it. "Was I *impatient?*" he once bellowed at a visitor, full of remembered anger twenty years after the fact. "Of course I was impatient! As a result of this *stupid*—excuse me, that's exactly what I mean—*stupid* reluctance to accept my vaccine, people were being crippled! People were dying!"

It did not seem to matter what he said. In the United States, the debate went on. The Salk vaccine had saved thousands of lives and had dramatically reduced the number of polio cases from an average of 16,000 a year between 1951 and 1955 to 4,500 a year in the next five years. But after five years of use, Sabin and his advocates pointed out, Salk vaccine had not eliminated the disease in the United States, and worse, some of the polio victims had previously been given Salk vaccine. But Salk and his followers maintained that to give a vaccine of live viruses was to sow the seeds for disaster.

The two became bitter antagonists, and before the issue was settled they confronted each other many times. Once, before television cameras, Sabin informed Salk: "Two thousand cases of paralytic polio in the United States, if they can be prevented, are two thousand cases too many. There is *no* 'irreducible minimum,' sir. I disagree with *that*." Salk seemed momentarily at a loss for words, although he continued to claim long after the issue had been settled that his arch-rival's vaccine represented an insidious danger.

By the summer of 1960 more than 100 million people worldwide had been immunized with Sabin's vaccine, and in countries where it was widely used, epidemics had immediately ended.

In August the Public Health Service, ever more cautious after the early troubles with the Salk vaccine, finally approved Sabin's vaccine, but commercial production did not begin until 1961. In much of the world, Sabin's vaccine was already used exclusively, and wherever it wasn't in use, he went and lobbied for it. Although East Germany had live vaccine, West Germany still used killed vaccine and they were not about to have a "Communist vaccine"—until an epidemic broke out and their vaccine could not stop it. In Japan the story of reluctance was similar; a change was not made until an epidemic in 1961 struck children who had received Salk vaccine earlier that year. In Italy, Sabin reprimanded officials who for obscure reasons had warehouses full of live vaccine that was not being distributed, although there were thousands of cases of polio in Italy. A new minister of health organized a nationwide vaccination program with live vaccine, and polio was quickly eliminated from the country.

In the United States powerful interests had opposed the change to oral vaccine: Basil O'Connor, the National Foundation of Infantile Paralysis, and eminent scientists had staked their reputations on killed vaccine. There were, too, the interests of the vaccine manufacturers. Killed vaccine was much more profitable than oral vaccine, which cost only pennies per dose. Further, it was not known what new difficulties and lawsuits might befall the makers of a new vaccine.

"To say that I fought the pressure to keep the vaccine out of use like an amoeba at the bottom of the ocean would be wrong," Sabin said. "I sometimes behaved more like a tiger, but I used facts. I used data. And the point was that I *had* to do it, because until 1961, I didn't have an organization to back me. Jonas Salk had an organization, the National Foundation of Infantile Paralysis."

It seemed to Sabin almost miraculous that his vaccine prevailed, though in retrospect he thought that maybe it had taken so long to win approval for good reason. It had been necessary to establish beyond any doubt that there was no danger from a vaccine that tampered not only with individuals but with the very ecology itself.

In June 1961, following a study by a special committee, the American Medical Association recommended that everyone, regardless of how many doses of Salk vaccine they had received, be vaccinated with the live vaccine. A physician in Phoenix, Arizona, Dr. Richard Johns, developed a program that helped accomplish this called "Sabin on Sundays."

Jonas Salk maintained that his vaccine was protecting people against Sabin's vaccine, that if his had not been given first, Sabin's would have caused calamity. But once licensed, Albert Sabin's live vaccine came to be used exclusively in the United States, as it already was in most of the world.

In 1955 the Soviet Union reported 17,000 cases of polio; in the same year, 27,000 cases were reported in other European countries, and more than 30,000 were reported in the United States, Canada, Australia, and New Zealand—more than 76,000 cases all together. By 1967, following the introduction of Sabin's live vaccine in these countries, the total number of cases declined to 1,000—an extraordinary reduction of almost 99 percent in a dozen years.

This is not to say that polio has been conquered. As later outbreaks of polio among unvaccinated English children and the Amish of Pennsylvania so plainly showed, the virulent viruses are still out there, awaiting the opportunity to rush in. But after Sabin's vaccine had been in use for five years, the number of cases in the United States dropped to less than twenty a year, and by 1975 the disease had been put completely under control.

It was one of the great medical triumphs of the century.

4
Anatomy of a Cold

Day One

THE VICTIM STEPS into an empty elevator. Mistakenly, she presses the lobby button and then the "5" for her floor.

It was the same elevator the night watchman had been riding seconds earlier on his way home from work. On the ride down he had been singing when he was interrupted by a sneeze. He cursed the cold spring weather and sneezed again, covering his mouth with his hand. Impatiently, he punched the lobby button again.

Walking along the hall to her office, the victim pulls a wind-blown strand of hair from the corner of her eye, using her right index finger and thumb.

Using an ordinary microscope, no one would detect anything unusual about the victim's finger, or the lobby button, or the air in the elevator—just the several million drops of moisture spewed into the air by the guard's singing and sneezing. No one would see the billions of infinitesimally small specks of "dust" clinging to each drop.

When the victim caught her breath in the elevator, a quarter of a million of these moist, dust-laden particles lodged in the minute hills and valleys of her nose and began drifting along the membrane lining toward her throat. Her fingers conveyed a second invasion from the elevator button to her eye.

Like undulating seaweed, invisibly wafting, the cilia lining the victim's nose and throat sweep away intruders at a rate of six hundred sweeps per minute. But the massive invasion of particles overwhelms them, and the dust entrenches itself in the warm moist terrain.

Drifting randomly, some of the dust particles bump into cells of the nose's membrane lining. The attraction is irresistible, and the particles cling to these cells.

Minute, twenty-sided polyhedra resembling geodesic domes, these particles are rhino, or "nose" viruses. So overwhelming is the attraction of the rhinoviruses for the cells of the nose that they ignore all others in favor of them. If they were rabies viruses, they would attack brain cells; if they were influenza viruses, they would fasten onto lung cells.

Very soon, hundreds of thousands of viruses are glued to the surface of the cells in the nose. Feeling the virus clinging to its skin, these cells pucker up, surround the rhinovirus, and drink it in. Influenza viruses—even more boldly—blast a hole in the cell membrane and fire their contents inside.

Having admitted a visitor to its inner sanctum, the host cell errs again and obligingly dissolves the virus's coat of armor with an enzyme intended to kill. In doing so it unleashes the virus's deadly essence: viral nucleic acid, which will direct the virus's reproduction.

Once undressed, the virus is ready to begin its reproductive rites. Lacking the essential machinery and materials for this, it rapidly wrests control of the cell's manufacturing center, subverting its processes so the cell will produce new viruses rather than cellular materials. The takeover, executed as strategically as a spaceship lift-off, starts with the firing up of viral genes that locate and shut down the cell's factory.

By the end of the day, the manufacturing centers in as many as 900,000 of the victim's nose cells are completely shut down. A second battery of viral genes fires up, directing cell-factory production of viral nucleic acid.

Day Two

When sufficient supplies of nucleic acid have been made, a third series of viral genes activates, switching production to proteins for the virus's armored coat. Twenty-four different types of amino acids are needed for the coat the virus will wear when it emerges into the world, and the cell is "instructed" how to manufacture those.

In hours, production work is complete and assembly of the young viruses begins in the cells. There is nucleic acid at the center of each new virus, and surrounding it, the protein-coat molecules in a tight triangular arrangement, forming the new twenty-sided polyhedron. The progeny, as many as one hundred in each cell, assemble naturally: it is the path of least resistance.

As early as twenty-four hours after the viruses invaded the nose, the new generation of viruses make their escape, massing toward the edges of the cells and bursting in a silent fireworks through the membranes. Fatally wounded, the cells will soon die. Within the hour, viral progeny—90 million strong—from the 900,000 cells drift toward healthy cells.

Arriving at her office in the morning, the victim coughs and clears her throat. "A sore throat?" she wonders and decides to have tea with lemon instead of coffee. The sensation disappears after a few swallows, but she tosses down some vitamin C tablets anyway.

In her throat, the viruses have been facing a tough battle. When she coughs, thousands land on her desk, and under the bright light, they survive only minutes. Others are washed into her stomach on floods of tea and water, where they are instantly devoured by acid.

Another contingent of viruses, drifting along, runs into trouble in the tonsils and adenoids, where it sets off an alarm in the victim's immune system. Patrolling lymphocytes, responding to the alarm, attack and kill many of the viruses and transport others back to lymph-node headquarters for scrutiny. Observing the pattern on the invader's coat of armor, the immune system begins producing antibodies that will seek out and kill the viruses. But there are far more of the enemy than the lymphocytes can rout, and several thousand viruses gain a hold.

Day Three

Some of the first 90 million viruses manufactured by the occupied cells have invaded other healthy cells and reproduced, and a second generation of viral progeny—9 billion of them—floods into circulation. Dead nose cells begin to accumulate by the millions, and the body secretes fluid to wash them from the battlefield. The victim experiences a slightly runny nose.

Day Four

A third generation of viruses pours into the victim's system. There are now close to a trillion, and for every virus the lymphocytes manage to kill, 10,000 new ones take its place.

The victim awakes feeling miserable. Her eyes and nose are rivers, her head aches, her throat is dry and sore. She calls in sick, huddles in bed, drinks orange juice, and takes aspirin and vitamin C tablets.

The viruses continue to proliferate and cell losses rise even higher.

Day Five

Something changes. The virus abruptly stops infecting new cells, and the body begins to wash away dead cells and rhinoviruses in great numbers. The victim feels better and returns to work, although she suffers cold symptoms for the next several days.

It is not known why, instead of persisting for weeks or months, the viruses stop invading cells and the cold ends. Antibodies, of course, limit or prevent the spread of a cold, but antibody levels have not been observed to rise significantly until *after* a cold has peaked. The viruses—rhinoviruses—which cause most of the billion colds in the United States each year and the loss of an estimated 40 million workdays and 30 million school days annually, have been studied intensively. There are more than one hundred varieties of rhinovirus, and antibiotics are not effective against any of them.

The existence of so many cold viruses means that a vaccine cannot be made. Since antibodies to one strain do not prevent infection by another, protection would require a huge vaccine of all one hundred viruses, and such a vaccine would be too dilute (there would be too little of each virus) to stimulate the formation of antibodies.

It would not be practical to vaccinate people against several strains of rhinovirus at a time until overall immunity is built up because resistance to colds, unlike many other viral diseases, is short-lived, lasting only a year or two. Nor is it possible to vaccinate against a prevalent strain, as with influenza, because several different rhinoviruses may be prevalent at once, and new strains quickly appear and disappear. Moreover, other families of viruses can also cause colds, and there are even some colds for which no virus can be found.

None of this was suspected in 1946 when Christopher Howard Andrewes, an English doctor, launched his all-out assault against the common cold. The "cold virus," as it was called then, was a vexing entity about which much less was known than most other viruses. That is, most scientists believed the agent was a virus (since it passed through filters), but no one had been able to capture it. The few attempts to discover the origin of colds, the causes, and how they spread had ended in frustration and confusion.

Because the virus could not be isolated, there was no way to study it or even ascertain its presence, except as it revealed itself in humans and chimpanzees. So a primitive test had been devised: secretions were washed from the throat and nose of a cold sufferer, strained through membranes too fine to let bacteria through, and dropped in the noses of chimpanzees or human volunteers. If colds appeared, it meant the secretions were infectious. But the yield was erratic: the results were often not reproducible, and sometimes no colds appeared—and no one could explain why.

When Christopher Andrewes informed the director of England's National Institute for Medical Research that he intended to take on the common cold, the director had temporized: "Well, you've got courage."

Andrewes had stumbled into viral research completely by luck, and bad luck at that. As an adolescent he had suddenly started to grow several inches at a time and his circulatory system was temporarily left behind, causing attacks of faintness and weakness. In winter, when the condition was worse, he was confined to bed, and he resorted to watching wildlife from the window of his home just outside London and writing down what he saw.

At school Andrewes had been assigned to classics, "a fate that befell anyone in those days who was of above average intelligence." But after he won first prize in an essay contest sponsored by the Royal Society for the Protection of Birds for his observations on how the feet and legs of certain birds are adapted to their function, the headmaster switched Andrewes to science. Andrewes was so happy the gamesmaster found it necessary to write in his reports: "He would do well to pay more attention to the game and less to the vegetation at the periphery of the field."

The success of this essay led Andrewes to consider a writing career, but his father, Sir Frederick Andrewes, advised him to obtain a medical education first so he would have his bread and butter. His experiences as a student at St. Bartholomew's Hospital in London, where his father was a professor, made Andrewes forget about writing. His father had helped document the devastating effects of the great influenza pandemic of 1918, and another professor there was enthusiastic about virus research. Andrewes was attracted to viruses because they were so mysterious and troublesome.

After qualifying in medicine, he went to work at the hospital for a researcher who believed he had found a virus that caused cancer. For Andrewes the work meant endless hours waiting at the operating theater to collect samples of tumors from the surgeons.

Andrewes eventually decided he was getting nowhere by trying to pursue both clinical medicine and research, so he joined the National Institute for Medical Research in Hampstead, just outside of London. In the laboratory where Andrewes worked, dog distemper was being studied. A highly contagious viral respiratory infection, distemper had been chosen for study in the hope that it might give insight into human respiratory viruses, and dogs had been singled out with the idea that if some relief could be afforded them, it might help rebut the arguments of the antivivisectionists. But the distemper work had run aground almost at the start because so many dogs were immune to the ailment.

It seemed to Andrewes pure luck that the head of the virus division, "a huntin' fishin' shootin' type" named Captain Douglas, happened to know that on farms and estates ferrets caught distemper from dogs, and had suggested using ferrets for the distemper research instead of dogs. In the laboratory ferrets readily caught distemper.

Shortly, toward the end of 1932, influenza broke out in London. During the course of the epidemic, Andrewes and another worker from the laboratory, Wilson Smith, attended a scientific meeting in London. There they met a professor who happened to remark that the epidemic offered a fine opportunity for someone to find the agent responsible. The year before the virus of swine influenza had been found by an American, Richard Shope. He had taken pigs' respiratory secretions, filtered them, and transmitted the disease to other pigs by nasal inoculation. The work was widely discussed because it suggested a way to proceed with the investigation of human influenza.

Wilson Smith was as much a pessimist as Andrewes was an optimist. But even Smith thought it might be worth trying to isolate the agent of influenza. He and Andrewes proposed to inoculate rats, mice, rabbits, and guinea pigs with washings from human influenza. Back at Hampstead, as they discussed their plans, Andrewes felt his temperature and pulse rise. Smith washed out Andrewes's nose and throat, and Andrewes went home to bed—with the flu.

Smith filtered the washings and inoculated the animals, but nothing came of it. Not long after that, the head of the laboratory, P. P. Laidlaw, received a telephone call from another laboratory, where the entire staff was catching the flu. To his surprise, he was told that all their ferrets, which were used for distemper research, were getting influenza too. Laidlaw mentioned this to Smith, who seized the clue. He took Andrewes's remaining throat washings out of the icebox and inoculated two ferrets that happened to be on hand for distemper studies.

Two days later—the day Andrewes returned to work—the ferrets began sneezing and looking watery-eyed. Their temperatures rose, the tips of their noses paled, their noses became stuffy, they yawned and refused food. To the great delight of Smith and Andrewes, the illness lasted several days, and they thought it looked as though the ferrets had caught the flu.

It turned out that the ferrets at the other laboratory didn't have influenza after all, but an unusual form of distemper. Smith and Andrewes's ferrets soon caught distemper as well, and between the flu and distemper, the two researchers were quite confused. Luck rescued them after Wilson Smith became infected, apparently from a ferret. Andrewes washed out Smith's nose and throat, and Smith went home to bed with the flu. Ferrets inoculated with his washings, which they called the WS strain, also caught the "feverish snuffling disease," and Smith and Andrewes thought the disease almost certainly was human influenza.

To rule out any doubt, Andrewes and Smith established a ferret hospital several miles from the main Hampstead laboratory where the ferrets could be kept in strict isolation. Every morning Smith, Andrewes, and Laidlaw drove to the hospital, donned high rubber boots, gauntlet gloves, and raincoats, swabbed themselves thoroughly with Lysol, and waded through a Lysol pool before visiting each pair of ferrets in their isolation cubicles. The precautions helped, but occasional accidental infections still slowed down their work.

When the washings were diluted a hundred times, they remained infectious—and inoculated into more ferrets, they consistently produced the flulike disease. Further evidence that the WS strain was influenza was the way it spread among the animals. Another sneeze helped clinch the case. After a ferret sneezed in the face of a new worker, Charles Stuart-Harris, he came down with the flu.

By the summer of 1933 the researchers were certain of their results and announced the news: they had found a virus for human influenza.

After that, "things went merrily," as Andrewes mildly put it. Their isolation of the virus led to the development of an influenza vaccine. The efforts of many went into this vaccine, which was made of viruses grown in hens' eggs and inactivated with formaldehyde. By the outbreak of the Second World War, there were several experimental vaccines in existence, but fears of another epidemic like that of 1918—which had started among troops—led to the rapid development of a vaccine suitable for large-scale use. This was done by Dr. Thomas Francis of the University of Michigan. After tests in 1942 and 1943, it was put into use by the mili-

tary. There were no major epidemics, and the vaccine was pronounced a great success.

During the war Andrewes worked on typhus, which is caused
by relatives of free-living bacteria, and in the process he contracted a severe case of the disease. By the time he recovered,
three years later, the war was over and typhus was being controlled with insecticide, so Andrewes began casting around for
new fields to conquer.

After the war, things had stopped going merrily with influenza with the alarming discovery that the virus could transform
itself! This was a potentially insuperable blow, in that the effec-
tiveness of a vaccine depends on how closely the vaccine-induced
antibody—like a lock—fits the keylike surface antigens of the
invader. The influenza virus had been transformed to such an
extent that the vaccine was rendered not just less effective, but
utterly useless against the new strain.

Similar antigenic changes in the influenza virus had been
charted almost simultaneously in distant parts of the world, and
Andrewes realized that international collaboration would be
necessary to cope with the problem. He suggested to the World
Health Organization that a network of influenza surveillance centers be established, and he set up the first one in England. Similar
centers were subsequently established around the world for the
purpose of isolating and comparing new strains that might surface. In theory at least, this would make it possible to prepare
and distribute a vaccine containing the new strains in time to
head off a major epidemic.

Andrewes believed that this was probably the most that could
be done at the time about such devious, changeable viruses, and
his thoughts returned to a project he had started years before—
the common cold.

In 1931, Andrewes's father suffered a stroke while in the
United States on business, and Andrewes and his mother had

gone to New York to be with him. While Sir Frederick recuperated, Andrewes visited a researcher at the College of Physicians and Surgeons, Dr. Alphonse Dochez, who was studying colds. By means of filtered nasal washings, Dochez had transmitted colds to humans and chimpanzees, and he believed that what he was transmitting was a virus. Dochez had also attempted to culture the agent in minced chick embryo, with peculiar results. After fifteen subcultures were done under anaerobic conditions— conditions not likely to keep cells, not to mention a virus, alive (but this was not then understood)—he produced colds in human volunteers, though none in chimpanzees, which are very susceptible to colds.

Andrewes thought the work interesting, and when he returned to England he tried to repeat the experiments. Chimpanzees were prohibitively expensive, so he used human volunteers. "We cannot get hold of any chimpanzees," he announced to a group of medical students at St. Bart's, "and the next best thing to a chimpanzee is a Bart's student." A hundred students volunteered.

Andrewes could not isolate his test subjects as Dochez had isolated his chimps, so he inoculated them in groups of eight or more. To some he administered active filtrates from colds; others served as controls and received a placebo. The experiment was seemingly a success: the medical students caught colds in numbers comparable to Dochez's chimpanzees.

Andrewes and a collaborator next tried to culture the filtrates, but with the cultured substance not a single student caught a cold. So Dochez came over to see what the problem was, bringing cultures with him. When Dochez and Andrewes tested more students with these cultures, only an insignificant number came down with colds, and Andrewes published a report titled "The Common Cold Wins the First Round."

By 1946 no one had been able to confirm Dochez's results, and the agent of colds was as much a riddle as ever. Andrewes thought he would take another try at "cracking the nut." He was optimistic that it would not be cracked, like influenza, only to reveal more nuts.

Andrewes's first objective was to establish a reliable method of detecting the presence of a cold virus. Since only humans and chimpanzees were known to catch colds and chimps were too expensive, this meant human volunteers would have to do until a better system was developed. Once he had a dependable method of detecting and transmitting colds, Andrewes intended to hunt for a convenient laboratory animal, isolate the agent, culture it, study it, and perhaps find a preventive or cure.

He began to look for a place where human volunteers could be kept in strict isolation. Experimenting with humans outside an institutional setting was a novel idea, and he did not know quite what he was looking for. The Ministry of Health urged him to use a wartime facility Britain had inherited from the Allies: the Harvard Hospital and a large complex of buildings just outside of Salisbury, ninety miles southwest of London. Constructed with funds from the Red Cross and Harvard Medical School, the facility had been designed as headquarters for epidemiologic teams and an infectious-disease treatment center in the event that epidemics had resulted from crowding in air-raid shelters. The feared epidemics never developed, and the unit was used for a time by the U.S. Army as a medical laboratory for its European forces.

Andrewes originally pictured the cold experiments on a modest scale, with a few volunteers involved at a time, but once he began thinking seriously about it, he realized that large numbers of subjects would be needed and that the Salisbury unit would be ideal.

The Americans had thought big: the Salisbury unit consisted of twenty-two buildings surrounded by sixteen acres of lush farmland and woods. Some of the structures were in poor condition, and Andrewes had them torn down; others were altered. He chose to use the oblong huts that had served as hospital staff

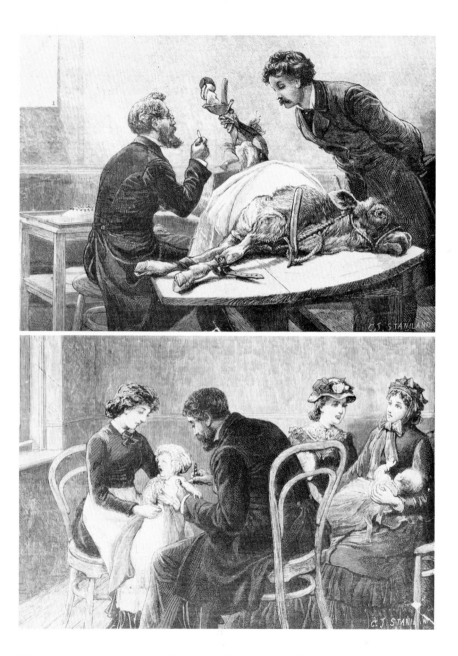

The first vaccine for a viral disease, smallpox, was a gift
from nature. "Vaccination from the Calf," wood engraved illustration from *The Graphic*, 1883.
(*National Library of Medicine*)

Louis Pasteur observes as his new rabies vaccine is administered by an assistant. Wood engraved illustration from *Harper's Weekly*, 1885. (*National Library of Medicine*)

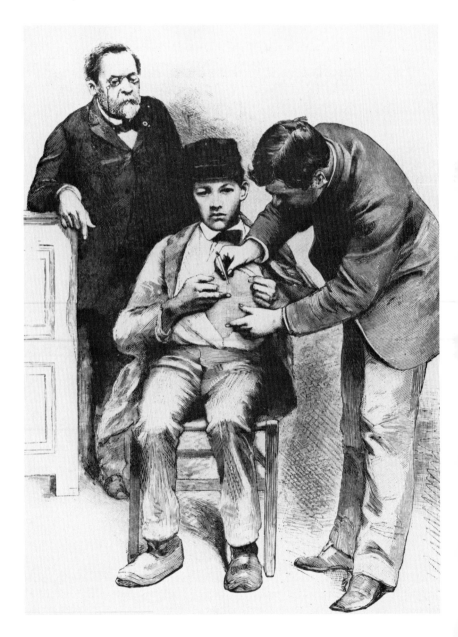

Resistance to the concept of vaccination vanished along with the smallpox epidemics. This etching by James Gillray, "The Cow Pock, or the Wonderful Effects of the New Inoculation," dated 1802, caricatures Edward Jenner giving his vaccine. (*National Library of Medicine*)

Camp Lazear in Quemados, Cuba, where Reed's experiments were conducted. (*Armed Forces Institute of Technology*)

Walter Reed, who led the Yellow Fever
Commission to Cuba in 1900. The commission proved how yellow fever was
spread. (*National Library of Medicine*)

Clara Maass, a nurse, volunteered twice to be bitten by mosquitoes. She died of yellow fever in 1901. (*National Library of Medicine*)

Max Theiler in Liberia with the Harvard African Expedition, 1927. (*Elizabeth Theiler Martin*)

Asibi. Researchers encountered him in 1927 in a Gold Coast village. He was suffering from yellow fever, but recovered in several days and went back to work. (*The Rockefeller Foundation*)

Max Theiler's final remark about the yellow fever vaccine: "There is no explanation." Theiler in his laboratory the year he won the Nobel Prize, 1951.

"Look," an old friend advised Albert Sabin. "Why don't you throw this stuff down the sewer and forget it?" In the end it seemed almost miraculous to Sabin that his polio vaccine prevailed. Albert Sabin in his laboratory at the University of Cincinnati, about 1955. (*March of Dimes Birth Defects Foundation*)

Albert Sabin's daughters in 1956: Debbe, six, and Amy, four. If he were not prepared to give the vaccine to his own children first, Sabin knew he could never give it to anyone else's. (*Albert Sabin*)

quarters to house the volunteers. Each hut, measuring a generous 20 by 140 feet, was divided into two apartments, with a sitting room, bedrooms, a bath, a kitchen, and a clinical room. Covered walkways were constructed between the buildings to make clinical visits and food delivery easier in bad weather.

Andrewes was pleased with the site, finding it pleasantly "airy," though at times it was too airy, being open to strong winds from the southwest. (A gale once blew the roof off one of the buildings.) The terrain was also pleasing, with its partridges, rabbits, and wild orchids, which thrived in the chalky soil. All in all, Andrewes thought the place ought to put the volunteers into an agreeable frame of mind.

Besides Andrewes, who became the first director of the Salisbury Common Cold Research Unit, there was a physician, an administrator, a matron to look after the volunteers, a secretary, a cook, a chauffeur-mechanic, and a succession of "girl graduates" who were trained as laboratory technicians, and who, among other duties, accompanied the director on his early morning walks to study the local flora and fauna.

Inspiration for the first experiments came from four Americans who had been trying to cultivate the agent of atypical pneumonia in eggs, without success. But when the egg substance itself was placed into the noses of hamsters and rats, the researchers achieved their goal. Andrewes decided to apply this approach to colds.

First, volunteers were needed, and obtaining them proved to be difficult. In the summer of 1946, he and other members of the Salisbury staff visited universities, gave lectures, and held press conferences describing the project and appealing for volunteers. It was at one of these press conferences that the unit acquired its famous reputation as "the place for honeymooners." Andrewes had been explaining that the volunteers would not be totally isolated, but housed in twos, when, as he put it, "a little devil caused me to say, entirely on the spur of the moment: 'It might suit honeymoon couples.' I shall never forget the sight of all those journalists pricking up their ears and then starting to scribble madly on their note pads," he wrote later in his book, *Pursuit of the Common Cold*. The name stuck, and it did not do any harm. The appeals always drew some response, and with experience the staff became more adept at enticing volunteers to spend their vacations in bucolic solitude at Salisbury.

Since there were few precedents for human experimentation, Andrewes devised a system as he went along. In return for a ten-day stay at the unit, volunteers received free room, board, pocket money, fare to and from Salisbury, and a bottle of beer, cider, or stout each day of their stay. Andrewes wanted only true volunteers; he did not feel right about experimenting on the retarded or prisoners, and he was grateful for the advantage of dealing with a minor illness, one people were not afraid of catching. Considering those who would serve as controls and those who would be resistant, most of the volunteers could be assured of not even catching a cold.

Many of the volunteers were students; there were also teachers, nurses, homemakers, miners, civil servants, gardeners, and factory workers. It became almost a patriotic duty to volunteer for a stay at Salisbury. Foreigners also volunteered, mainly from Holland, Scandinavia, the United States, and Australia.

All that was required of the volunteers was that they be between eighteen and fifty years of age, not pregnant, and not suffering from a cold or other infection. Once a burglar came (only once, Andrewes said), and once a man who said he had been called a sex maniac but that it was untrue. (He was disqualified.) Now and then there were difficulties, in remarkable variety, with the volunteers, as one of the physicians, Dr. A. T. Roden, complained:

> Ten little volunteers
> On the Salisbury line.
> One of them rode through to Bath
> Then there were nine.

Nine little volunteers
Coming through the gate.
One of them had second thoughts
And then there were eight.

Eight little volunteers
Made the number even.
One of them looked *very* odd
Then there were seven.

Seven little volunteers
Up to monkey tricks.
A fire bucket fell on one
Then there were six.

Six little volunteers
Hoping to survive.
An X-ray shadow settled one
Then there were five.

Five little volunteers
Peeping round the door.
One had a nasty cold . . .

Andrewes enjoyed the volunteers; he had never cared for the
ferrets, which bit and had to be grasped tightly by the neck before being picked up. Most of the volunteers cooperated, and
some returned many times. By 1970, more than 17,000 had visited
the unit. If the rest of his scheme had unfolded half as smoothly,
Andrewes would have rejoiced.

On arrival the volunteers met, were given lunch, a pep talk,
and a physical examination. Once an experiment began they were
rigorously isolated: the matron and physician wore sterile masks
and gowns when visiting them, and they were not permitted to
go into town, ride in cars or buses, go within thirty feet of
others, or talk with anyone they might accidentally meet. When
hot meals were delivered to their rooms, they had to wait a few

minutes before opening the door, so as not to break their isolation.

Each apartment was equipped with a radio and telephone, newspapers were delivered each morning, and volunteers were free to roam the countryside. If they played badminton, volley-ball, bumblepuppy, or any of the other games provided, they could play only with their living-quarters partner; afterward, they opened the windows and cleaned the equipment with disin-fectant, and the next pair to use the room had to wait an hour before entering.

After several days of isolation, when the physician was certain that none of the volunteers had colds, inoculations were given. All the tests were double-blind: some of the volunteers received cold viruses, others a harmless saline solution, and neither physi-cian nor subjects knew which until the results were tabulated. A virologist in the laboratory coded the information. If the con-trols came down with colds, all the results of that experiment were discarded.

On daily rounds the doctor inquired about sneezing, watery eyes, stuffy nose, aches, sore throats, cough, headache, malaise, chills, fever, shivering, lassitude, irritability, and other cold symp-toms. Each symptom was scored from zero to nine and the num-ber of points totaled. Five types of colds were defined, ranging from abortive (slight respiratory symptoms that are gone in twenty-four hours) to severe (requiring bed rest and accom-panied by shivering, aches, fever, and more serious respiratory symptoms).

Some of the volunteers, as if they thought it was expected, provided elaborate, exaggerated reports of their symptoms, while others insisted they felt splendid, even as they coughed and sniffled and shivered. This difference was taken into account.

A good index to a cold, Andrewes perceived early on, was the number of tissues used by a volunteer. The tissues were saved and counted both before and after inoculation, and the ways different people used a tissue was also taken into account: some blew their noses once, folded it neatly, and threw it away; others used a tissue until it was wet and shredded.

"All this may sound rather unscientific," Andrewes allowed, "but in practice it worked well," and he soon attained his first objective: a dependable method of transmitting colds. Using a boundless supply of cold viruses from the boys at Harrow School and Salisbury workers, the staff inoculated volunteers and found that they could reliably produce colds in thirty-five to forty percent of cases. This, in turn, provided a means of determining whether they had found a culture technique that worked.

They had not. They first tried cultivating the agent in hens' eggs, the technique that worked with influenza, herpes, and other viruses. Nasal washings from cold sufferers were injected into the membrane, yolk sac, amniotic sac, and other parts of the chick embryo—all with negative results. When the "cultured" viruses were inoculated into volunteers, they did not catch colds.

Reports came from the United States that a cold virus cultured in eggs had produced colds in volunteers. So Andrewes visited one of these laboratories, in Baltimore, and returned home with some of their culture viruses. "Alas," said Andrewes, who used the word often in connection with colds, "in Salisbury they produced no colds at all."

More culture fluid was sent over from the United States, but it didn't produce colds the second time either. Then the Baltimore researcher brought some over himself, and under his supervision it caused colds. After he left, Andrewes and the staff repeated the experiment—with no results. They were stymied. Andrewes concluded that the volunteers used in the Baltimore studies, who were prisoners, probably told the doctors whatever they wanted to hear. Other laboratories also failed to reproduce the Baltimore results.

Such work continued for years at Salisbury. It was tedious and amounted only to disproving the claims of others.

The hunt for a convenient laboratory animal susceptible to

colds was equally exasperating. Ferrets refused to catch colds. Capuchin monkeys from South America, which have a watery-eyed appearance—and, Andrewes learned belatedly, are known as "weeper monkeys"—were said to catch colds, but at Salisbury they did not. Neither did South African green monkeys or other species of monkeys, flying squirrels, grey squirrels, cotton rats, guinea pigs, hamsters, voles, hedgehogs, cats, baboons, sooty mangabeys, or pigs. But Andrewes, ever the optimist, found something to be cheerful about. Postwar rationing was still in effect, but since the pigs had been used for experimentations, they could not be turned in to the rationing board, and they made delicious eating.

Failing to culture the agent or transmit it to a laboratory animal, Andrewes began to concentrate on determining the virus's physical properties, which might enable him to guess which family of viruses it belonged to and provide a lead. The researchers found that, like other viruses, the cold agent could survive for up to two years at $-76°$C., and was inactivated by heat. It appeared to be inactivated by ether, but when this experiment was repeated, ether did not seem to kill it.

As for the virus's size, it sometimes passed through filters with very small pores, and at other times it passed through only much larger filters.

Andrewes wondered whether, because of its differing characteristics, the common cold virus was actually several or many different viruses. A newly discovered family of viruses, the adenoviruses, supposedly caused coldlike illnesses, but in tests at Salisbury, these viruses caused more severe symptoms than the common cold Andrewes was pursuing. Another incomprehensible clue! After years of work, Andrewes realized that all he had learned was how to give someone a cold.

To ward off feelings of frustration, Andrewes embarked on an investigation of how colds spread and invited members of the National Institute of Medical Research team on air hygiene to join in. The ever-accommodating Salisbury volunteers this time put their heads into large plastic bags and breathed, coughed, sneezed, or recited Shakespeare, as they chose. The thousands of particles they emitted in doing so were collected and counted. But most of the airborne particles were heavy and fell quickly to the ground—so it was unlikely that they were the main source of transmission for the disease.

Andrewes wondered if handkerchiefs might be the cause of trouble, so the researchers collected handkerchiefs from volunteers with colds. The secretions were washed off, centrifuged to concentrate them, and deposited on fresh handkerchiefs. "Alas," Andrewes sighed again, "volunteers who used these handkerchiefs did not catch colds."

Perversely, the Salisbury volunteers did not seem to catch colds from one another, either. When cold sufferers were paired with healthy volunteers, there was one cold in nineteen cases at first, then one cold in five, then none in four.

The experimenters grew more desperately ingenious. A Salisbury researcher attached a gadget to his nose that emitted fluid at about the rate of a moderate cold, and he used a handkerchief to blow his "nose" as necessary. "The fluid contained a dye normally hardly visible but fluorescing brilliantly in ultraviolet rays," Andrewes later said. "He spent some hours in a room with other people playing cards, eating a meal, and so on. At the end of the time, the lights were turned off and a U-V lamp revealed the horrible truth; his artificial nose secretion had got around everywhere—all over his face and clothes, his food, the playing cards."

Since this experiment pointed to a different conclusion from all the others, Andrewes deemed that contaminated hands, along with handkerchiefs and other inanimate objects, were "unlikely to be very important."

The error of this deduction passed unnoticed by anyone for many years.

As the negative results accumulated, Andrewes's determination grew by bounds. His next plan, the abandoned island experiment, was so ambitious that all his previous experiments paled by comparison. Isolated individuals, such as sea travelers, Arctic explorers, and entire isolated communities, had long been known to be completely free of colds, until they came into contact with the outside world. Then they became ultrasensitive to colds. Thwarted at Salisbury in learning how colds spread, Andrewes created his own isolated community after hearing of a deserted island called Eilean na Roan (the isle of seals) off the northern coast of Scotland.

The tiny island, barely a mile long and a mile across, was situated in a region where colds were rare and greatly feared, for when they occurred they were severe. The island and its dozen houses had been abandoned in 1938 because of the inhospitable land; only sheep grazed there.

Andrewes persuaded nine students from Aberdeen University to spend several months on the island, and twenty tons of equipment were packed in boxes, transported by boat to the island, and carried up its steep cliffs. From July until September 1950, the students lived isolated from the rest of the world except for radio contact. In mid-September, when Andrewes and Salisbury researchers arrived, they inoculated some of the students with a cold virus. Once again they encountered the curse of cold research: the inoculum was no good, and not a single cold developed in the island inhabitants either after inoculation or person-to-person contact with the outsiders. It was, Andrewes allowed, "an astounding piece of bad luck." On the bright side, he added, "It was one of the most glamorous experiments in which I have ever been engaged, for the Aurora borealis was throughout displaying overhead."

All told, a tremendous amount of work had been done since 1946—and the results were pitiful. They could not culture the cold virus; they had not found a laboratory animal with which to experiment; they had not learned how colds were transmitted; and Andrewes had not achieved his main goal: the discovery of a laboratory test to recognize the commonest kind of cold virus without using human volunteers.

When the unit was reevaluated for funding by the Medical Research Council, it was nearly closed down; only Andrewes's irrepressible enthusiasm kept it going. It was not easy to make a case for the unit, he recalled. "It was costing a lot of money and we hadn't provided a cure for the common cold."

A nice, solid clue finally turned up in an epidemiologic study of families living in the region. Families with schoolchildren, it was found, had many more colds than others, and the children themselves had more colds than adults, indicating that schoolchildren were an important source of colds. But an experimental children's birthday party held at Salisbury, attended by children and adults, did not shed any light on how children might spread colds.

Andrewes was desperate, so he lowered his sights and took on the very modest project of examining some of the age-old myths about colds. He designed a "rather harsh" experiment to get at the most sacred adage of them all: chilling causes colds. Diluted cold washings—enough to produce only a couple of colds—were dropped into the noses of six volunteers. Six other volunteers took hot baths and then stood in a cool hallway for half an hour, wearing wet swimsuits. Then they dressed but wore wet socks. Another six volunteers were subjected to both the virus and the chilling routine.

The outcome was that two volunteers who had received the virus alone caught colds; no one in the chilled group got a cold; and four in the chilled, virus-infected group caught colds. "So far so good," Andrewes said, "but the numbers were small." When the experiment was repeated, the opposite occurred. The virus alone produced more colds than the double treatment of virus and chilling. In another experiment, when the volunteers were thoroughly chilled by walking in the rain and returned to cold rooms, there was no indication that chilling had made them sensitive to the virus; and all succeeding experiments on chilling continued to be negative.

"All the same, I think something happens in the autumn that upsets your balance and gives the common cold a chance," Andrewes still believes. "Some kind of upset, something to do with the weather plays a part." If some sort of stress is necessary to lower resistance, then it is not surprising that these experiments failed. The Salisbury volunteers were probably not sufficiently stressed. As a matter of fact, they found the experiments quite amusing.

After a practical method of culturing many viruses in monkey-kidney tissue was devised in Boston in the late 1940s, the Salisbury staff redoubled their efforts to culture the cold agent. A Brazilian, Dr. H. G. Pereira, came to direct the work, and he tried growing cold washings in nasal epithelium and in human embryos, which were obtained from Sweden. At first Pereira did not have any luck: of 76 volunteers inoculated with cultured viruses, only 2 got colds. But early in 1953 something seemed to change. Washings from a member of the Salisbury staff cultured in series produced colds in 13 out of 120 volunteers after as many as ten subcultures. It was not many and there was no observable change in the cultures, but it was enough to suggest that the virus multiplied. It was a heartening development, and it happened at a propitious time because the Medical Research Council was again reevaluating the unit for funding.

Funding was renewed, but then there was another setback. The cultures did not produce another cold, and the original batch was soon gone. Pereira and the staff worked for three years to find the problem, varying factor after factor in the culture system. They tried embryonic trachea, monkey tissues, and human embryonic kidney, all to no avail.

In 1957 a new virologist, Dr. David Tyrrell, arrived and began testing all the variables in a complex series of permutations and combinations. He soon showed that viruses nurtured at 33°C. produced more colds in volunteers than those grown at 36° or 37°C. In retrospect it was an obvious point, since cold viruses do not incubate in the bloodstream but in the membrane lining of the nose, where it is slightly cooler.

Working with three different strains, which Tyrrell grew at the cooler temperature in embryonic kidney cultures, the Salisbury staff was able to transmit colds much more regularly to volunteers by 1960. The substance they used had been cultured as many as eight times, and they were confident that something was multiplying in the cultures, even though it could not be seen.

Tyrrell then used the haemagglutination test, which is based on a virus's ability to cause red blood cells to agglutinate or mass together, and on the ability of one virus to interfere with the growth of another kind of virus. To cultures that were believed to contain cold viruses, Tyrrell added a test virus, parainfluenza, which caused red blood cells to agglutinate and thus provided an indirect means of measuring the presence of other viruses. In these cultures, much less haemagglutination was caused by parainfluenza virus than in the parainfluenza virus controls, which contained no cold viruses. In other words, there were cold viruses present in the first set of cultures, which prevented the parainfluenza viruses from multiplying.

"It was the high-spot of the whole research!" Andrewes said. It was what he had set out to find in 1946: a way to detect the presence of cold viruses in the laboratory without using volunteers. No matter that it had taken fourteen years; no matter that in 1933 a similar accomplishment with influenza—the transmittal to ferrets—had taken about fourteen days. Here it was!

The culture technique was refined: the viruses liked a dash of bovine plasma albumin, a little more glucose, a little less bicarbonate of soda, and gentle rocking as they incubated, which washed the cells with nutrients and removed wastes. Under these improved conditions the virus flourished. (It was thought that the first success had been due to an embryo that was especially good, and that once it was gone, they had been unable to grow the viruses again.)

In the meantime, cold viruses had been isolated in the United States at the Naval Medical Research Unit #4 in Illinois as well as in other laboratories, and it was realized that an important new family of viruses had been found. Sir Christopher Andrewes, as he was by then, christened them rhinoviruses, or nose viruses, after their preferred habitat.

Researchers eagerly applied to come to Salisbury, and investigations into immunity, vaccines, treatments, and cures were undertaken. Vaccines were made and tested at Salisbury and elsewhere; but no sooner had hopes for the conquest of colds been raised than they withered. By 1967, 55 different antigenic types of rhinovirus had been found; the number soon rose to 80, then to more than 100, and there may be still more. Indeed, almost every fact that has been uncovered about the rhinoviruses suggests that the common cold will *not* be conquered.

The answer to one of the most puzzling questions pursued by Sir Christopher Andrewes—how colds spread—came in 1977 from an investigation at the University of Virginia School of Medicine. Earlier, it had been established that when rhinoviruses were placed directly in the mouth or throat, infection did not readily occur, but that the eye was easily infected. But how were they entering the eye? Continued attempts to recover rhinoviruses in significant numbers from the air, sneezes, or coughs had failed. Reviewing the facts, Dr. Jack Gwaltney at the University of Virginia was as puzzled as anyone. Then, in studying colds among insurance-company employees, Gwaltney and his collaborator, Dr. J. Owen Hendley, made the discovery that rhinovirus infections did not spread as easily at the workplace as they did in the home. This indicated to them that some sort of close contact was necessary for the efficient spread of the viruses.

At about this time, the accidental spread of colds in their laboratory from infected to noninfected volunteers during nasal examinations led Gwaltney and Hendley to formulate a theory for the route of infection: contamination of a cold sufferer's hands with their nasal secretions, followed by hand contact with another person, followed by accidental self-inoculation at the nose or eye by the contaminated fingers. Gwaltney and Hendley observed that rubbing the eyes and placing the fingers in the nose are natural behaviors for adults and children alike—occurring as frequently at Sunday schools as at medical-society meetings. In adults they observed one finger-eye and one finger-nose contact every three hours.

Back in their laboratory rhinoviruses were easily transferred from the hands of cold-infected volunteers to those of non-infected volunteers during a short ten-second contact, and in the recipients, brief hand-to-nose contact did lead to colds. Gwaltney and Hendley also found, as the Salisbury workers had, that the air is an inefficient vector for these viruses. They placed volunteers in chairs around a small table, and those with colds coughed, sneezed, sang, and talked loudly, but only one of twelve susceptible volunteers caught a cold. Gwaltney and Hendley tried to gather rhinoviruses directly from volunteers who coughed and sneezed into petri dishes containing a collecting broth—also unsuccessfully. Similar attempts by researchers at Fort Detrick and at the National Institute of Allergy and Infectious Diseases in Maryland were also largely ineffective. A few colds are transmitted by airborne particles, but the hands clearly had it.

The facts defy good sense, of course, since sneezes and coughs have always been closely associated with colds, and Gwaltney and Hendley wondered if these symptoms might just be another of nature's deceptions. If, as William Budd, the British epidemiologist, said: "It is not often that nature wears her heart on her sleeve or delivers up her secret at the first summons. Quite as often it seems to be her mood to mislead by deceiving shows."

To this day most of what can be said about colds is what we do not know. We can only guess at how many different cold

viruses there are. Colds and coldlike illnesses are most common
in the winter, when people huddle indoors, and in the fall an
increase in colds might be explained by children returning to
school, exchanging colds, and passing them on to adults. But
why aren't there more colds in the summer? It may be that
temperature and humidity in the spring and fall, when rhinovirus
infections peak, favor the virus or weaken resistance or both.
But this is only conjecture.

It is not known why a cold makes a person feel so uncom-
fortable or how or why a cold sometimes causes a fever. There
is only a small amount of inflammation and swelling of the nasal
passages in proportion to the symptoms; and it may be that the
symptoms are not caused by the virus itself but by the body's immune response, though this too is only a guess. As for fever, it
has been suggested that the body's rise in temperature may inhibit growth of the virus.

Some people with frequent colds have a structural difficulty,
such as clogged eustachian tubes or sinus problems, that may
worsen a cold, or they may sleep with their mouths open, al-
lowing the membrane linings to become dry and thus more susceptible. People who travel tend to catch more than their share
of colds because they encounter more different strains. Teachers,
doctors, theater managers, and others who have frequent contact with people are also prone to colds. But there are some
colds for which no explanation—no structural problem, no allergy, and no virus—can be found.

The role played by stress is not understood either; nor is it
known what stresses, if any, might turn a harmless encounter
with a virus into an infection. Naval recruits have been found to
catch a lot of colds during training; and the number of colds
was reduced by delaying their standard immunizations and re-
laxing the general discipline. But how such stresses might operate is unclear. There is no scientific evidence, either, that
becoming run-down is a factor in catching colds. The answers
to all of these questions have been pursued, but not found.

And, of course, no one knows how to prevent, treat, or shorten the length of a cold. Modern cold remedies are no more effective than those of Hippocrates's day, when bleeding was prescribed (he disapproved), or those of the first century, when Pliny the Younger recommended "kissing the hairy muzzle of a mouse." Commonly used antibiotics are ineffective against rhinoviruses and all other viruses, and well-run tests have found antihistamines of little value: they may alleviate symptoms but they do not cure. Most other nonprescription remedies, on which an estimated $750 million a year is spent in the United States, do not even ameliorate symptoms, according to the Food and Drug Administration. The effect of aspirin on a cold has not been determined, cold sufferers who take aspirin excrete viruses longer than those who do not, but the significance of this is not understood.

Drinking fluids and bed rest do not affect the course of a cold, although they may make the victim more comfortable. Dr. Francis Lowell of Harvard University, who directed a Food and Drug Administration study of cough and cold medicines, concluded that "chicken soup is as good as anything." A National Institutes of Health physician who specializes in colds once remarked: "In my line I get a lot of colds. I treat them by doing nothing. I don't take any drugs—antihistamines make me sleepy and aspirin upsets my stomach."

Careful tests at the Salisbury Common Cold Research Unit on the efficacy of vitamin C showed that it had no observable effect on rhinoviruses, or any other viruses tested. Twenty-nine volunteers were given three grams of vitamin C a day, beginning the third day before inoculation with cold viruses, and twenty-six volunteers were given a placebo; in each group there were nine colds. Only one difference between the two groups was reported: volunteers who were told before the end of an experiment that they had received vitamin C maintained that it decreased the length of their colds. When volunteers were not so informed, both the vitamin C and control groups reported the length of their colds to be the same.

Sir Christopher Andrewes points out that some of Dr. Linus Pauling's famous experiments demonstrating the purported benefits of vitamin C were not controlled. "Pauling said his experiments did not *need* controls," Andrewes says, rolling his eyes skyward. "If you don't have controls, you get badly led up the garden path."

For his part, Dr. Pauling, the Nobel Prize-winning physicist, maintains that the colds induced at Salisbury differ from true wild colds. But the fact remains: there is no good evidence that vitamin C is effective against rhinovirus infections. It must be noted, however, that the problem may simply be the lack of an experimental method to demonstrate an effect, rather than the lack of an effect. Vitamin C may work, but we may not know how to prove it.

Antiviral drugs have been tried against colds, and here, too, difficulties abound. Because viruses insinuate themselves so intimately into the cellular machinery, a good antiviral agent must make the ever-so-subtle distinction between virus and host cell and destroy one without harming the other. Since colds are rarely serious (except in babies, smokers, and those with lung ailments, in whom colds may trigger bronchitis or asthma), any antiviral agent used must also be beyond a shadow of suspicion and have no side effects. Further, the short incubation period for colds means that by the time a victim is aware of symptoms, the infection is likely to be at its peak.

Whether interferon, the antiviral protein manufactured by the body, will be useful remains to be seen. Interferon seems to protect against all viral infections, in contrast to antibodies, which protect only against specific diseases. It is species specific, however—mouse interferon, for instance, does not work in humans—and until human interferon was made in the laboratory in 1980, there was little hope of obtaining enough to study it intensively (65,000 pints of human blood were needed to extract 100 milligrams of interferon). With gene-splicing techniques it is expected that enough interferon will be produced to test its potential benefits.

"It is said that if as much money were put into common cold research as was used to put men on the moon," Sir Christopher Andrewes wrote once, "the problem could be solved. Things aren't, however, as simple as that." The fact is that after thirty-five years of effort, the lowly cold viruses remain as troublesome as ever. Andrewes is inclined, however, to exult in how much has been learned by studying viruses—an outlook that may be attributed partly to his lifelong custom of not putting all his eggs in one basket. After retiring as director of the Salisbury unit, Andrewes wrote ten books on viruses and entomology, and having decided that that was enough writing, he continued to collect bugs.

He thought it great good fortune that he was possessed with the collecting instinct (his father had been interested in entomology and an uncle was an expert on Indian beetles). He has found insect hunting the finest of hobbies: "It brings intellectual interest as well as physical benefit; it provides an objective for outdoor exercise, it is relatively inexpensive and it furnishes, if one collects, some occupation for the winter evenings and other times when fieldwork is impossible." It is a hobby that has never let him down.

A large, sturdy-looking man at the age of eighty-three, with a florid outdoor complexion and a fluff of undisciplined white hairs, Andrewes rested one damp November morning in the sitting room of his cottage, Overchalke, which perches at the upper edge of an old chalk pit just outside of Salisbury. At Andrewes's elbow were a tea tray, binoculars, and two insect-eating plants, a sarracenia and a Venus's-flytrap.

Very much the fond parent whose infant has just taken its first solid food, he reported that the flytrap, a scrawny wisp of greenery only a couple of inches high, had already consumed several wasps. At that moment, a cough from his guest caused Sir Christopher to ask if he might be in danger of catching something. When told that it was a recent case of pneumonia, he inquired if the organism had been identified and an expression of relief crossed his face on hearing that it was haemophilus influenza, an antibiotic-sensitive bacterium, rather than a virus.

As tea was poured by Lady Andrewes, Sir Christopher regarded his visitor with one eye. The other was covered with a gauze bandage, inflamed with a case of shingles. This was the result of a return visit by the virus that causes chicken pox. The virus, also called herpes zoster, hides out quietly in the nervous system and reappears years later when something—no one knows what—upsets it, to cause a painful irritation of the sensory nerves.

Despite a severe case that put him in the hospital and laid him low for months, the irony of this had not escaped Andrewes. Having devoted the better part of half a century to the pursuit of viruses, he had decided to leave viruses well enough alone— but they refused to do likewise. Rather than provoking rancor in their victim, however, the attack by the herpes virus had quite the contrary effect: Andrewes still admired viruses' unfailing ability to surprise him. Nevertheless, he has come to prefer insects.

At first he chased butterflies and moths, Andrewes explained, but they were already so well studied that there was little to discover, so he turned to bees and wasps. He wrote a book about them and is an authority on diptera, a large order that encompasses many flies, mosquitoes, and gnats. Some of his specimens may be found in the British Natural History Museum, and the remainder crowd his small study at Overchalke. His virology texts and papers have been relegated to one corner, and hundreds of shallow wooden boxes are stacked on shelves from floor to ceiling. In the boxes, one may glimpse long, even lines of specimens, hundreds upon hundreds of them, all neatly skewered with straight pins, labeled, and pinioned once and for all in place.

5
The Virus That Ate Cannibals

Washington, D.C., March 1957

DR. JOSEPH SMADEL PICKED UP a thick, airmail envelope from the pile of mail on his desk and studied the stamps. There was only one person likely to be writing to him from New Guinea, and Dr. Smadel pondered how remarkable it was that from ten thousand miles away a person could make such a nuisance of himself. Carleton Gajdusek seemed to cause more trouble now than when they had worked together under the same roof.

By the age of ten Carleton Gajdusek had made up his mind to become a scientist like his aunt rather than a businessman like his father. His father, a Slovak farm boy who had arrived in the United States alone and unable to speak English, had become a prosperous butcher, and Gajdusek was raised in a large house on a hill overlooking the crowded immigrant community of Yonkers, New York, in the 1930s. "Our family had 'risen,'" he said once.

He and his brother grew up listening to his mother, who was university educated, read aloud from Homer, Virgil, Hesiod, and

Sophocles. With his entomologist aunt, he searched the country-side for small plants and animals. They cut open strange lumps on plants, discovering insects inside, and collected twigs with gummy masses on them that hatched indoors, covering the curtains with praying mantises.

Carleton read Paul de Kruif's *Microbe Hunters* and inscribed the names of the twelve microbiologists on the steps leading to his chemistry laboratory in the attic. At about the same time he decided that he would study chemistry, physics, and mathematics rather than biology in order to prepare for a career in modern medicine. He loved school, although he got into trouble for bringing jars containing insect-killing potassium cyanide to class. Carleton seemed to succeed easily at anything he did. One summer, working in his aunt's laboratory, he had set out to synthesize an insecticide and instead produced a new compound that became a patented weedkiller. "The only venture I have had which involved commerce," he later remarked.

At seventeen Carleton entered college, and by the age of nineteen he was off to medical school at Harvard. He went on to specialize in pediatrics and neurology, completing his training while at the same time conducting research in physical chemistry at Caltech with Linus Pauling and then in virology with John Enders at Harvard. His professors called him "Atom-Bomb" Gajdusek, because he was always exploding with ideas.

As a junior member of Dr. Smadel's department of virus and rickettsial diseases at the Walter Reed Army Medical Center in Washington, Gajdusek was full of plans, but from the beginning he seemed to lack discipline and was constantly launching into overly ambitious field studies that took him to remote places. When he abruptly departed for the Middle East without a word of explanation to anyone, Dr. Smadel felt he had no choice but to assume that Gajdusek did not intend to do any further work for the army.

Plans were made to terminate Gajdusek's salary, but he proceeded to send back to Washington reams of data and drafts of articles on epidemics in the Middle East and elsewhere—enough information for ten scientific papers. As a result, Gajdusek's salary was extended even though his writings were, as Dr. Smadel expressed it, "a goddamn mess." Smadel spent hours working over the material. In one case he removed lengthy digressions on the geography and culture of the Middle East only to find them inserted in the revised draft of another paper that had been sent to Gajdusek for approval. The next time, Dr. Smadel instructed him, "do not add the innumerable conversationally interesting points about the travelogue aspects of the trip. . . . So what if they live in mud houses or holes in the ground." By the time all the papers were revised, Dr. Smadel was thoroughly exasperated and informed Gajdusek that in the future he should stay at home and prepare his own papers.

Several years later, a letter arrived from Gajdusek who was in Melbourne, Australia, working in a laboratory on hepatitis immunity. He wrote that he was on the trail of a significant discovery. But his grant was about to expire, and he was looking for a new position that would enable him to continue the project.

In his spare time, Gajdusek went on, he was studying the primitive inhabitants of Queensland and the islands of New Guinea and New Britain off the northern coast of Australia, and he had begun a number of child-growth and development studies there—but only as a hobby. "I am still interested in chasing the Indians in the Amazon Basin," Gajdusek told his old mentor, who had left Walter Reed and was associate director of the National Institutes of Health. "But as far as wanting to roam continually, year-in-year-out—NO! I am not physically up to it. . . ."

Unlikely, Dr. Smadel thought. It was hard to imagine Carleton Gajdusek's feeling frail. He was a tireless worker, and although he could be stubborn and unpredictable, he always came up with worthwhile results. Dr. Smadel decided that Gajdusek would be an asset to the National Institutes of Health and wrote back telling him so.

Gajdusek replied that he would return to the United States by the summer of 1957, ready for full-time work at the National Institutes of Health. Meanwhile, he was busy night and day with his hepatitis research, hurrying to complete it and publish his findings before senior members of the laboratory—who had become very interested in his project—swooped down on it.

The latest letter, which Dr. Smadel now pondered, was indeed from Gajdusek, who gave his address as the Okapa Patrol Post. By way of explanation he had added "Forei, Kimi, Kei-agana Linguistic Areas. Territory of Papua and New Guinea." The letter, dated March 15, 1957, read in part:

Dear Dr. Smadel,

I am in one of the most remote, recently opened regions of New Guinea (in the Eastern Highlands) in the center of tribal groups of cannibals, only contacted in the last ten years and controlled for five years—still spearing each other as of a few days ago, and cooking and feeding the children the body of a kuru case (the disease I am studying) only a few weeks ago.

This is a sorcery-induced disease, according to the local populace, and that it has been the major disease problem of the region as well as a social problem for the past five years, is certain. It is so astonishing an illness that clinical description can only be read with skepticism, and I was highly skeptical until two days ago, when I arrived and began to see the cases on every side . . . a mighty strange syndrome.

To see whole groups of well nourished healthy young adults dancing about, with . . . tremors which look far more hysterical than organic, is a real sight. And to see them, however, regularly progress to neurological degeneration in three to six months, usually three, and to death is another matter and cannot be shrugged off. . . .

What he saw could not be senility, Gajdusek continued, since young and old people alike were affected by it. He thought it could be a psychological or genetic disease on an epidemic scale. If so, it was extraordinary. Whether of infectious, toxic, or allergic origin, it was remarkable, he thought, and ought to interest everyone in medicine.

Gajdusek wrote that he intended to get to the bottom of the mystery, but he needed help. Since the first of the year, he had been without a salary, and his own savings would last only a few more months. If Dr. Smadel could be of assistance, Gajdusek could promise a series of papers on the new disease when he returned home, and he mentioned that he had work pending in a dozen other fields as well.

Even if his friend could not help, Gajdusek intended to stick it out. "This," he concluded excitedly, "is nothing to be put off and nothing to be dropped!"

In Papua New Guinea, a rough Jeep track cut through the jungle linked the territory of the Fore people to the nearest town, Kainantu, forty miles to the northeast. In good weather it was a hard day's journey; in rainy weather the route was impassable due to landslides.

Venturing into the region for the first time, Gajdusek had scarcely been able to believe his eyes. The Fore men, with pigs' tusks in their noses and their dark bodies glistening with pig grease, were an awesome spectacle. For ages a fierce mountain jungle had shrouded their Stone Age society from the outside world, and these cannibalistic tribes, living in huts hidden in deep grass on the lower mountain slopes and in the rainforest at higher altitudes, had begun to come under governmental control only in the past five years. Even more remarkable than the warriors were some of the villagers, who with their absurd smiles, shrieks, and shivering seemed to be out of their heads.

As Gajdusek saw more of the strange ones in villages remote from each other—all behaving in exactly the same bizarre fashion—he began to think there might be more to their behavior than eccentricity or hysteria. The fact that within a short time this behavior gave way to serious neurologic symptoms and then to very real death convinced him that there was.

The Public Health physician for the region, Vincent Zigas, had written to the Walter and Eliza Hall Institute of Medical Research in Melbourne, where Gajdusek was working, about the phenomenon, but no one other than Gajdusek had expressed much interest in his discovery. When Gajdusek arrived to see it for himself, Vin Zigas eagerly told him everything he knew about it.

A Lithuanian who had emigrated to New Guinea seven years before, Vin Zigas had first noticed the syndrome among the natives two years earlier. He told Gajdusek that it seemed to be spreading.

Together, Gajdusek and Zigas made an expedition into Fore territory to hunt for cases. Most of those Zigas had examined the previous year were dead, villagers told them, but there were newly afflicted people in every village. Using interpreters, pidgin English, and pantomime, the two physicians interviewed victims and their families. Almost immediately, Gajdusek noted a startling similarity in their accounts. The victims always became sick gradually, and unsteadiness was one of the first insidious signs. Within a month, shivering and difficulty with speaking set in, and in three months the sufferers were almost totally debilitated.

As more people were interviewed in the next days, Gajdusek observed that the shivering disease—descriptively named *kuru*, to shiver, by natives—often struck the same family more than once, sometimes years apart. In one family he learned that the mother had been dead for five years, an older sister for three years, and now a ten-year-old girl was showing the first ominous signs: a vacant expression and clumsiness when she walked.

The victims, several in each village they visited, were a heartbreaking sight. Only months before, according to natives, they had appeared completely normal. Gajdusek saw them trembling, unable to stand, barely able to speak, and unable to feed them-

selves. Many were near death. Since no one was able to do anything for them, their plight was utterly hopeless.

The natives' acceptance of their fate only made the situation
more heartrending. In the earliest stages of the disease they
laughed good-naturedly at their own clumsiness, even though
they knew full well what was in store for them. As they became
incapacitated, they continued to accept their plight with equa-
nimity and grace. Families gave strong emotional support to the
kuru victims to the very end—brothers and sisters sleeping with
their arms around the ill one, parents cuddling a sick child, a
husband lying patiently beside his unmoving, stinking wife. The
vengeful hunt for the kuru sorcerer, Gajdusek thought, also
must have been a source of comfort to the dying.

Like all diseases, kuru was attributed by the natives to sorcery,
which was worked on its victims in retribution for some insult,
such as rebuffing a suitor. Something belonging to the intended
victim was taken—the skin of a potato the victim had eaten, or
clothing, for example—and an elaborate rite ensued. The charm
was wrapped in leaves and bark and buried in a swamp. Each
day the sorcerer visited the site and dug up and shook the charm
until the victim, too, began to shake.

Revenge on the clan guilty of the sorcery was had by *tukabu;*
they were ambushed and then viciously pounded with stones,
bitten, and their genitals crushed. Gajdusek marveled that some
of these victims survived, though he observed that while most
of the kuru victims were women, all the sorcerers were men,
and *tukabu* served the useful purpose of helping to restore the
ratio of men to women.

The Fore community was in chaos with ritual murders taking
place everywhere; these vendettas caused nearly as many deaths
as kuru did. If this state of affairs continued, it wouldn't be
long before the Fore people were wiped out.

Gajdusek quickly became convinced that kuru was not some
morbid cultural oddity but a serious medical problem—the most

bizarre and baffling one he had ever seen. Determined to get to the bottom of it, he formulated a plan: he would plot the location of all known kuru cases, study them, locate new cases, and attempt to treat the ill. Being a pediatrician, he would concentrate on the children, hoping to increase his chances for success. His training in pediatrics was considerable, and he believed there were few childhood diseases about which he had not at least heard.

It was a logical approach and in another setting would have been easily implemented. But here the obstacles were enormous. The Fore lived in hundreds of small hamlets scattered over some of the steepest, roughest terrain in the world, and most could be reached only by walking; they spoke a language Gajdusek was just beginning to comprehend; and there was no one to help him except Vin Zigas, when he could get away from his regular duties. As Public Health officer and the only physician in the entire Eastern Highlands, Zigas was responsible for the care of the European population in the town of Kainantu.

Gajdusek made his headquarters in the hut of the government patrol officer for the Fore region, the Okapa Patrol Post, and immediately started scouting for cases of kuru. He covered the nearby region by Jeep and began to travel further afield on foot. He listened to the victims' stories and collected blood samples, which he sent for analysis to the Hall Institute in Melbourne and to Dr. Smadel in Washington. He asked the Hall Institute if he might send them the brain of a kuru victim, providing he could obtain one, as well as other specimens for virus-culture work. He also arranged to send a young kuru victim to Melbourne for examination and, hopefully, treatment. She was in the early stages of the disease, could still travel easily, and was cooperative. An older boy who could translate would accompany her, and her family was willing. The only difficulty was persuading the girl to wear clothing.

The government patrol officer, Jack Baker, helped Gajdusek hunt for kuru cases when he had time, but he was busy dealing with the ritual murders that the Fore committed upon one another as revenge for kuru sorcery, so Gajdusek was often the only nonnative for miles around. Fortunately, the Fore were friendly and cooperative, and he was allowed to live in their villages to hunt for kuru cases. Once word had spread about the white doctor, sufferers were brought to him.

As far as Gajdusek could tell, there had been no kuru sorcery until about twenty years earlier when it appeared in a few villages before spreading to others nearby. It could appear at any time of year, natives told him, and it struck mostly women and children. Wherever Gajdusek went, natives acted out the distinctive tremors, leaving little doubt in his mind that they were all describing the same affliction. He could not locate any victims who remembered being very sick, as with encephalitis, before kuru set in, which was medically unprecedented.

Vin Zigas wrangled some supplies and a microscope from the Public Health Department, and Gajdusek began performing lumbar punctures—on the patrol officer's dining table—on any kuru victims who could be persuaded to submit to the procedure. Examining the spinal fluid in fourteen cases, he found nothing abnormal, so he sent the specimens to Melbourne for further testing.

Within a few weeks of arriving in the Eastern Highlands, Gajdusek was busy all day and half the night trying to obtain supplies, patrolling the bush for kuru cases, collecting family genealogies, keeping charts on the victims, taking blood samples, and trying desperately to treat the sick. He divided his patients into three groups and gave antibiotics to one, vitamins and an improved diet to another, and small doses of phenobarbital to a third. Then he watched for any change in their condition.

There were also other illnesses to be treated, minor surgery to be done, and the confidence of the Fore to be won so that when a kuru patient died he might acquire the brain for autopsy. It was too much for one person to handle, so Gajdusek began recruiting and training some of the native boys who tagged curiously after him as his medical assistants and translators. When the patrol officer's house started overflowing with his kuru patients in late March, Gajdusek began to construct a kuru hospital with the assistance of the natives.

April 1957

By the beginning of April, Gajdusek and Zigas had located forty active kuru cases, and they had tracked down another forty who had died in the past year and a hundred more who had died in the past ten years of kuru.

With natives leading the way, Gajdusek trekked to the southern boundary of the Fore region and the edge of the neighboring Awa and KuKuKuKu territory to determine how far kuru extended. The expedition led him to think that the disease might be restricted to the Fore and their immediate neighbors, about 35,000 people living in 150 villages. Kuru seemed to be the major disease among them, and it was second only to warfare as a cause of death.

The kuru hospital, made of wood with walls of woven pitpit grass, was soon completed. Gajdusek surveyed this one tangible result of his efforts and pronounced it splendid. Perched at 6,500 feet, it overlooked vast valleys, and when the thick mists and clouds lifted, Mount Michael, towering 12,500 feet, was visible in the distance. The hospital was snug and dry during the storms that continually pounded down on the mountain top, there were mats on the floor for patients to lie on, an open fire for warmth, and considering the circumstances, it was reasonably well equipped.

It took time but it was not difficult to persuade kuru victims to stay at the hospital, and soon there were twenty-five, half of them children. But as the days passed, it began to look as if none of Gajdusek's three treatments were working. No patient showed any sign of improvement, and within a few weeks, they were all undeniably worse. Gajdusek had to admit that he was getting nowhere. He was certain only that the mystifying illness was a degenerative process of the central nervous system and that it progressed rapidly and relentlessly to a fatal conclusion. He couldn't treat it, and he did not have a clue to its cause.

There was evidence for genetic predisposition—in one family seven close relatives had died of kuru—although the victims' hysterical behavior made him think it could still turn out to be a psychosis. He wished there were someone to consult.

He and Zigas had written repeatedly to colleagues in Australia and the United States urging them to come and see kuru, but no one came. Despite this, Gajdusek remained convinced that kuru was important. Everything about it—the high incidence in such a small population, the strange symptoms, the speed with which it progressed—told him so.

He and Zigas decided to draft a short report on kuru while they were still the world experts, even though it would be far from conclusive. Gajdusek was beginning to worry that Zigas might be cheated out of credit for the discovery by his superiors at the Public Health Department—and that he himself, for that matter, might receive short shrift once other, more important scientists became involved. Surely, he thought, they would become interested any time now, and when they did there was bound to be the fierce jealousy and competition that inevitably swirls about any choice research project.

He didn't know, either, how much longer he could hold out in New Guinea. He desperately needed money to buy axes, beads, and tobacco to trade for dead bodies and permission to perform autopsies. His own money was quickly running out, and it was not clear how much longer the local Public Health Department would support the project or tolerate Vin Zigas's absence from his duties. Gajdusek calculated that if he set aside enough money to return home, there would be enough left for only two more months in the bush at most.

He wrote to Dr. Smadel again, asking if he agreed that kuru was important and, if so, would he help. Gajdusek said he was sure that if kuru could be cracked, it would offer a rare clue to the diseases of the nervous system, which were so little understood. He added, "I stake my entire medical reputation on this matter."

This time his request brought prompt results. Dr. Smadel replied that he had rounded up $1,000 for him, but with the news came a scolding: "Sometime, chum, you better settle down and learn how to look after yourself." If Gajdusek could manage to get himself back to Washington to sign a few papers, Dr. Smadel suggested, he would be put on the payroll at the National Institutes of Health, and that would save everyone a lot of trouble.

No sooner had word come that his money troubles were temporarily settled than trouble began brewing on another front. Sir Frank MacFarlane Burnet, the director of the Hall Institute in Melbourne, had learned that kuru specimens were being sent to the United States, and he began complaining that foreigners were stealing kuru. Gajdusek was furious and dispatched long letters to everyone concerned in Australia, the United States, and New Guinea, pointing out that he had repeatedly asked for collaboration in no fewer than five letters to the Hall Institute.

"An American working in an obviously 'choice' research project . . . was a decided fly in the ointment," he later said, "and for a matter of some weeks I was all but ordered out of the area —a suggestion I did not accept." Instead, he "only screamed for more collaborators and assistance," from the Hall Institute and the Public Health Department of New Guinea; and he got them.

Gajdusek continued to send specimens to Australia, but since Dr. Smadel was so helpful and efficient, Gajdusek began to rely on him for supplies, laboratory work, drugs to be tested, and words of encouragement.

As all the kuru patients continued to deteriorate, all the tests of liver, blood, urine, and spinal fluid continued to reveal nothing abnormal. He tried more drugs: aspirin; a drug used for multiple sclerosis; antibiotics; antiworm drugs; heavy metal detoxifiers; anticonvulsants; and hormones. Since grown men never seemed to get kuru, he tried testosterone. Again he watched anxiously for signs of improvement, but he was beginning to suspect that there wouldn't be any.

May 1957

The first hospital patient died, and at two o'clock one morning during a howling storm, Gajdusek set to work by lantern light. Using the only knives that were available and his bare hands, he sectioned the brain and worked late into the morning to perform a complete autopsy. The brain and tissues were fixed as best he could under the circumstances and sent off to Melbourne.

Gajdusek soon obtained a second brain and fixed it according to detailed instructions for jungle conditions that had been forwarded by Dr. Smadel. The first step was soaking the brain for several weeks in a solution of formaldehyde and salt. He then wrapped it in lint, drenched it with fixative solution, sealed it in cellophane, packed it in wool, and put it in a strong box for shipment. It was relayed by Jeep to bush pilots in Kainantu, who connected with commercial flights from Port Moresby to Sydney, Europe, and the United States.

The specimens were precious, and had been obtained with great effort, and Gajdusek did not know if he could get any more. He requested every study the pathologists in Melbourne and Washington would consider. For he was becoming ever more convinced that the secret to kuru must lie in its victims' brains.

Within weeks he decided not to do any more autopsies for the time being. He had pushed the natives as far as he dared: "They are proud and have their own ideas, which are most intelligent, and although they have conceded that I can cure their meningitis and pneumonia, they have decided that this magic is too strong for me and that my prolonging life . . . is no blessing at all. They want to die at home; and once fully incapacitated they want to die as quickly as possible."

He did not at all blame them, and he was deeply touched

when, "to humor me and repay my many miles of mountain climbing to track them down, they haul the litters over miles of cliff-faced and precipitous jungle slopes to bring the patients in for another shot at our therapeutic trials and experimental poking."

The simplest of procedures had been difficult before, but everything was becoming much more difficult, requiring more explanations, more bribes, more valuable time. Public opinion did not allow him to use catheters for collecting urine specimens, so he got up as many as half-a-dozen times in a night to collect specimens. And then smoke from the wood fires and debris from the grass-roofed huts, which were constantly blown everywhere by the high winds, contaminated most of them. Even so, the tests showed no signs of mercury, copper, or any other metal poisoning.

As Gajdusek was on the verge of ruling out metal poisoning, he considered other possibilities. He wondered if the cause of kuru could be manioc, a plant eaten by the Fore containing prussic acid in its raw form, which had been introduced into the area at about the time kuru first appeared; or something in the strong, raw ginger the Fore ate; or a substance in any of the hundreds of other foods they consumed.

Whatever it was, it was causing more and more cases of kuru. In some areas he visited, up to ten percent of the population was ill with it, and all the patients were declining in precisely the same way.

It was a truly astounding sight. Gajdusek worried that no one would believe him. He wrote to Dr. Smadel, requesting movie and tape-recording equipment and instructions on their use.

June 1957

A check for $1,000 arrived with a note from Dr. Smadel: "It is certain . . . that this particular check is the last one, so you better get back to the United States before you run out of money."

Ignoring this advice, Gajdusek set off to locate the western boundary of the kuru region with native boys preceding him to cut a path through the jungle and others carrying provisions. As he hiked and climbed, he was drenched, chilled, and badly bitten by insects. He was allergic to lice, but there was no way to avoid them, since it would mean rebuffing an affectionate hug or nose rubbing by the natives.

After trekking for twenty days, Gajdusek reached the border between the Fore and Kimi people and was rewarded, for here, he learned, kuru occurred only rarely. When it did, he noted that it could usually be traced to intermarriage: a Fore girl, for example, who had married into the Kimi caught kuru years later.

Gajdusek believed that most evidence pointed to a genetic cause for the disease. Environmental factors or diet could not be ruled out, but in light of this new information they seemed less likely. On the other hand, red herrings continued to turn up; genetics could not explain the case of a Kimi girl who caught kuru after coming to live in Fore lands. Never had he been so confounded!

By the end of June he had observed a hundred active kuru cases, and the victims were rapidly starting to die. The first neuropathology report arrived from Australia, but it contained no leads—only the information that there were possibly slightly fewer Purkinje cells in the cerebellum of the brain.

July 1957

More bad news: *Science* magazine rejected the kuru manuscript. "In its present form," the editor wrote, "it is very loosely written and rambling and quite unsuitable for publication." On top of that, the small amount of money that had been coming from the New Guinea Public Health Department (his only source of financial support aside from Dr. Smadel) ran out. Gajdusek was tremendously annoyed, because applying for another grant would take time, the one thing he—and the patients —did not have. Meanwhile, he would have to beg for supplies. He urgently needed a typewriter and thousands of file cards to keep up his kuru case histories.

If he couldn't crack kuru—and this was looking more likely every day—he was determined to at least obtain a complete record of every known kuru death before he left. It might provide a clue and would be vital to any future follow-up studies. So far, he and Vin Zigas had recorded 500 cases, but even if kuru were limited to the Fore people, the task was only half finished since a large territory remained to be surveyed.

At the end of the month a preliminary report arrived from the neuropathologists at the National Institutes of Health in Washington. They had found no sign of inflammation or infection—either bacterial or viral—in the kuru-afflicted brain. They did report finding significant degeneration in several parts of the brain and told Gajdusek that kuru was *definitely* an important new disease—unlike any other.

Gajdusek was jubilant: after five months of work this was the first concrete evidence that he was not off on a wild goose chase!

August 1957

Lucy Hamilton, a friend of Gajdusek's from the New Guinea Public Health Department and an expert on native nutrition, arrived and began helping Gajdusek search the soil, foods, and fire ashes of the Fore for traces of metal. The director of psychiatry from the Royal Melbourne Hospital paid a visit, examined kuru patients, and said it was definitely not a psychotic or hysterical disease.

The intensive hunt for the cause only uncovered more patients—there were 150 now, and 24 had died. Gajdusek knew he was far from being able to help the ill, and he had not even been able to get a report about kuru published. The editors of the *New England Journal of Medicine* were divided on whether to accept his and Zigas's report, owing to the lack of pathologic findings; so Gajdusek wrote to Dr. Smadel and asked him to assure the editors that kuru was not a figment of his imagination and that a pathologic report would be forthcoming.

Gajdusek was very discouraged; he was unaccustomed to failure, and it looked as if he had encountered his first big one. He was thirty-four years old, broke, and after six months of the hardest work of his life, he had a stack of case records on a disease he knew almost nothing about. Wherever he turned he saw grim testimony to his failure, and he was plagued by the thought that he ought to be doing *something* for the dying. But what?

And what was Gajdusek to advise the Public Health Department of New Guinea, which was trying to decide whether to isolate the Fore and let them die out, or to encourage them to outbreed and possibly risk spreading kuru throughout the world?

Although bands of native boys still tagged after him and gathered around, acrid and smoky smelling, resting their arms on his shoulders or rubbing him with pig fat, while he typed, Gajdusek knew patience among their elders was wearing thin. The Fore had willingly gone along with most of his peculiar ideas before, but now even his simplest requests met with excited, suspicious chatter, if not outright disapproval.

"Kuru has us licked," Gajdusek confessed in a letter to Dr. Smadel, although he vowed to finish delineating the kuru region. The most difficult part lay ahead: the lands to the south and east of the Okapa Patrol Post were forbidding, with river banks nearly as steep as the Grand Canyon dividing Fore territory from that of the KuKuKuKus, who were more belligerent than the Fore and were known as the Apaches of New Guinea. Gajdusek decided to fly over the region first, before beginning extensive explorations on foot.

Alarmed at reading these plans, Dr. Smadel wrote back immediately. He knew most lines of argument would be lost on Gajdusek once he made up his mind to do something, so Dr. Smadel urged him to consider the invaluable kuru records and all the information he carried in his head that would be lost if his plane went down in the jungle or the natives reverted to cannibalism. Since it was beginning to seem that Gajdusek might remain in the bush forever, Dr. Smadel had arranged for his appointment as a visiting scientist at the National Institutes of Health, effective immediately upon his return. Lest the point be lost on Gajdusek, Smadel added: "I think you better finish up your work cautiously . . . and get the hell back here."

At the end of this letter from Smadel came important news: "Additional histological sections have come through, and the boys are excited about the diffuse degenerative involvement which goes far beyond the cerebellum."

The pathologists had found a veritable battleground in the brain of Yabaiotu, a fifty-year-old female kuru victim. There were striking changes in the nerve cells of the cerebellum—the Purkinje cells—as well as other bizarre deformities, including numerous strange round or asteroid bodies and "torpedoes," a ballooning of the Purkinje-cell axons. Marked changes extended as far as the basal ganglia. The pathologists stated that the findings fit no known disease pattern.

There was still no sign of inflammation, they stated, and thus no reason to suspect an infection; but experts at the National Institutes of Health did not think it could be a hereditary disease either, because it developed too quickly and over too broad an age range. In short, they did not know what could have caused such massive destruction of the brain. The chief epidemiologist at the National Institute of Neurological Diseases and Blindness, Dr. Leonard Kurland, urged a more careful look at the food and water of the Fore, focusing both on its preparation and consumption. He also recommended that Gajdusek stop wasting time on further clinical descriptions of the disease and futile trial treatments. Others at NIH suggested that the causal agent might be a parasite or a toxin such as a cyanide-containing tapioca plant.

Gajdusek was amazed. The findings and the excitement of these authorities made him feel more energetic than he had been for weeks. With new confidence he again reassured Vin Zigas, Lucy Hamilton, his friend Lois Larkin who had come from Australia to help him, and Jack Baker, the patrol officer, that it was very unlikely that they could be accidentally infected with kuru as a result of the autopsy work.

By summoning every argument he could think of and talking for hours in each case, he managed to obtain another twelve kuru-afflicted brains, which he sent off to Washington. But he knew they would definitely be the last. A police patrol in the region had been attacked recently, and although the natives had not yet aimed any of their arrows at him, he knew he must be an ever more tempting target.

By the end of August he had located 200 active kuru cases, and 50 of the victims were dead.

September 1957

Gajdusek trekked up and down mountains, through tall grass and jungles, to hamlets in the far reaches of Fore territory, some never before visited by an outsider. He collected more kuru case histories, each entailing hours of discussion, and collected blood samples, which were relayed back to a kerosene icebox at Okapa by his loyal native assistants. On the journey he provided his helpers with sweet potatoes, corn, cabbage, and onions and paid them with tobacco, salt, and matches. Gajdusek also "overlooked their failings like an overproud father and defended them vigorously against any criticism." They knew this, and he was rewarded with their loyalty and affection.

While he was gone, the nutritionist Lucy Hamilton continued to scour for clues. She examined plants, animals, foods, medicines, and herbs of the Fore; she collected ashes and smoke from their cooking fires and samples of their body paint; she poked into their gardens and smoky, unventilated huts. The men and boys lived apart from the women and girls, who shared their dwellings with small children and the pigs; and she studied the living and sleeping habits of both groups.

Several promising leads resulted: the Fore consumed the leaves and bark of the Agara tree, which induced kurulike tremors, to gain insight into the future and see "dream-man." It was also discovered that women and children ate a small spider as they worked in the gardens. When a team of entomologists and an expert in Melanesian languages arrived from Australia, the hunt intensified. Hundreds of plants and insects—fleas, ticks, flies, mosquitoes, and spiders—were collected, tagged, and shipped off for laboratory study.

The end result of all this work was not a single lead. Patients taken out of the kuru region continued to deteriorate, and Fore people who had left their tribal lands long ago were suddenly stricken with the fatal disease. One entire village had fled across a river to escape from kuru, but the disease came with them.

When Gajdusek returned to camp to replenish his supplies before setting off on the last and most difficult leg of his survey, the only bright spot was a note from a chief neuropathologist at the National Institutes of Health. Dr. Igor Klatzo praised Gajdusek's "splendid work" and confirmed that kuru was unquestionably a new condition unlike anything in the medical literature. He mentioned that it did resemble Creutzfeldt-Jakob condition, a rare illness of which only twenty cases were known so far, but Creutzfeldt-Jakob affected a different part of the brain and it apparently did not strike children.

November 1957

On the fourth of November, Gajdusek returned to Okapa—finished. He had been accompanied by loyal native assistants all the way, and he insisted that his foray into the land of the Kuks sounded far more dangerous than it had actually been. Some years later, having on a subsequent visit to New Guinea lost a Jeep over a cliff during a landslide and having learned of the deaths of Michael Rockefeller, a doctor, and a patrol officer in Dutch New Guinea, he decided that New Guinea was a more dangerous place than he had realized in 1957.

Traveling by foot, Jeep, raft, canoe, and light plane, he had established definite geographical and cultural boundaries to the disease and plotted the trade routes and marriage patterns of its victims. It had been a huge task, and in a rare moment of satisfaction he congratulated himself: "Fifteen hundred to two thousand miles of some of the hardest walking in the world is now done!!!!" Furthermore, he knew the job had been done right. (In this he was proved correct, for teams of researchers who later retraced his steps found his epidemiology faultless.)

Back at camp he learned that the report on kuru had been accepted by the *New England Journal of Medicine* as well as the *Medical Journal of Australia* and *Klinische Wochenschrift*, the German medical journal. It would soon be in print on three continents!

Word about kuru had already spread, and eminent Australian professors visited the bush in droves. To Gajdusek's astonishment, they proclaimed that kuru was the most important neurologic discovery in decades and that Gajdusek was a hero. They were horrified at the thought that all his invaluable kuru records, stored in a grass hut, might be lost in a fire.

Their reactions confirmed his belief that kuru was a phenomenon that could not be ignored, and every bit as important as he had imagined. Buoyed, Gajdusek again began to think the kuru mystery might be solvable.

Reporters descended on Okapa, much to his irritation. He was not about to let anyone waste time that he could spend on kuru research. The journalists had taken to calling kuru "laughing death," and Gajdusek abhorred their sensationalistic accounts, which were appearing in the Australian press and on radio and television. Learning that reporters from *Time* and *Life* were on their way to see him, he sent word that he was heading deep into the bush, and he did not return to camp until they had given up and gone home.

Arriving back in camp, he was horrified to discover that his colleagues had been wasting time talking to the journalists and that the reporters had gained access to his confidential kuru files. "If it were not for the wonderful fun I have with these marvelous natives for whom I hope we may eventually bring some relief to their affliction," he fumed, "the bastards on the periphery of our problem would drive me from it." The publicity was especially embarrassing in that not a single word about kuru had appeared in any professional journal.

Gajdusek continued to weigh different theories—genetic, infectious, and toxic—briefly favoring one or another as a new finding turned up. But he kept returning to the idea that kuru must come from something the victims ate.

Back in the middle of September, a native had told him that his clansman had eaten his own grandfather, against the native's advice. Although cannibalism was on the decline, episodes of the cannibalistic rite, which was performed out of respect for the dead by family members, were not unusual. Gajdusek wondered if consumption of brain tissue might not cause long-term sensitization. "It is a wild idea, but evidently this disease is a wild one. . . ."

Yet as far as he could tell, cannibalism did not involve the brain, and it was most unlikely that all the kuru victims had participated in the rites; certainly the youngest had not. He decided that "cannibalism as a possible source seems well ruled out."

He wondered if there might be something in the undercooked pig meat the Fore ate, or in the leaves of the Kegata bush, which they ate along with pig and human flesh, or in the bark of the bush, which they sometimes chewed with another type of bark.

More samples were gathered and dispatched to the National Institutes of Health for toxicologic studies. He even sent along some of the stones used by the Fore in cooking. But the Fore ate well, and some of them were even fat; studying all the hundreds of foods they consumed would be almost impossible. He could not imagine anyone wanting to do it anyway.

Gajdusek knew he was grasping at straws; he couldn't even find a reasonable starting point for an infectious or toxicologic study. Out of leads and out of money, not to mention prospects of support for further fieldwork, Gajdusek reluctantly prepared to depart for home. He worked frantically: organizing studies to be continued in his absence, establishing a long-term kuru monitoring system, revising and correcting drafts of reports on kuru, and writing long letters on kuru matters to Australia and Washington. When he had the opportunity, he snatched as many kuru specimens as he dared—brains, visceral tissue, and "a good long piece of spinal cord."

Reel upon reel of film and dozens of shipments of autopsy tissue had already been dispatched to the National Institutes of Health, and he sent a final deluge. He was acutely aware that, "The natives have given up on our medicine. . . . They know damn well it does not work," and he was "fighting (verbal battles in Fore), bribing, cajoling, begging, pleading, and bargaining for every opportunity to see a patient, and strenuously working tongue muscles for hours for every further day we get a patient to stay in the hospital." He was thin and tired, and Vin Zigas was completely worn out. Suffering from a knee injury and nervous and physical exhaustion, Zigas was planning to take a year off to recuperate.

As Gajdusek had predicted, there were political storms swirling about kuru, and he found himself in the middle of them. The Australian medical establishment was arguing over who should step in; Adelaide University had proceeded full steam ahead with kuru field research, while both the Hall Institute and the Public Health Department of New Guinea maintained that kuru belonged to them. The accusation was constantly made, and continued to be made for years, that Gajdusek had stolen kuru. But as one observer put it, "Everyone concerned got scientific mileage out of it, and others would have gotten more if they had recognized the importance of it sooner. Gajdusek sent kuru specimens to the Australians, but they didn't move fast enough. They weren't accustomed to someone like Gajdusek. He wanted results, and he wanted them right away."

The Public Health Department was particularly distressed at the idea that Gajdusek might take his kuru field notes with him when he left. Gajdusek maintained that no one else could decipher his field notes, that it would take months to organize and copy them, and that they belonged to him in the first place. He had made carbon copies of the case records, which he intended to leave behind, and these had already been plundered by university students for their degree work. He felt that if anyone deserved to use the remaining material it was Vin Zigas, who did not yet have an M.D. Gajdusek thought that Zigas needed a bit of a push, so he planned to help Zigas prepare his thesis from this material back at the National Institutes of Health. Gajdusek had already asked Dr. Smadel to dig up a grant for Zigas. "Left to this pack of wolves," he said, "Vin would not long survive."

Letters arrived constantly from Washington, urging Gajdusek to hurry home. "It is good to learn you plan to return to the States in December," Dr. Smadel wrote. "I hope nothing comes up which will interfere with those plans."

December 1957

Australian scientists were due any day to take over the kuru field investigation, and Gajdusek wanted to be on hand to orient them, so he wrote to Dr. Smadel that his departure had been briefly postponed.

He now suspected that only an infectious agent could explain the fact that patients took kuru with them out of the area, and although he searched for clues until the last moment (arsenic in fingernail specimens? abnormal globulin levels in the blood?), Gajdusek told himself that if an answer were to be found anywhere it would be in the laboratory.

1958

In the middle of January, Gajdusek tore himself away from Okapa, although he promised his native friends to return soon. Dr. Smadel's relief at this was short-lived, for the end of the month found Gajdusek in Dutch New Guinea having "a fascinating time" and he had spotted "a half-dozen new problems—some good!!"

Gajdusek finally arrived home in April, traveling by way of Malaya, India, Russia, Finland, and Sweden. As he had promised, Dr. Smadel had a job waiting for him at the National Institute of Neurological Diseases and Blindness. It was unusual for an American to be a "visiting scientist," but that is what Gajdusek was for the time being; it had been the most expeditious way to get him on the payroll. He proceeded to design a project titled the "Study of Child Growth and Development and Disease Patterns in Primitive Cultures," which would allow him to pursue kuru and all his other interests at the same time.

In the National Institutes of Health laboratories, the kuru specimens were scrutinized from every angle, and every test the neuropathologists, biochemists, virologists, and bacteriologists could think of was done. It all came to nothing. Beyond the remarkable brain lesions there were no major new findings. Every attempt to locate an infectious agent failed.

By summertime Gajdusek was feeling footloose, and in August he was back in New Guinea, hiking through the bush, sitting around fires talking with his Fore friends, and hunting for clues. One chilly, rainy evening as he sat in a hut eating a plate of soup into which raindrops splashed from a hole in the roof, he realized that he felt as much at home there, surrounded by his Fore friends, as he did back in Washington—perhaps even more so. Gazing at a sad-eyed Melanesian, he felt a pang of his "puritanical conscience, an ill feeling at having been away from the work to which I should be devoted," and he chastised himself for a "lapse in laziness."

"I am more than ever ready," he felt, "to give myself to these wonderful Papuans who need our assistance so much and who have so much to teach us!" They filled him with admiration and sympathy, and he despised the attitude of some of the patrol officers, who regarded them as wretches. Gajdusek thought that their ritual cannibalism, which was practiced by close relatives of the deceased as a rite of respect, was not very different from crying at western funerals.

In an effort to repay the Papuans, who had shown him as much warmth as he had ever known and taught him so much, he began to help promising boys apply for scholarships and wrote them letters of recommendation. But as much as he loved the Fore, he also loved their fierce neighbors to the east, the KuKu-KuKus. They had an extremely good opinion of themselves—he saw something of himself reflected here—and while his Fore assistants willingly did their own work and that of the Kuks, the Kuks stood smugly by and watched them. He found them insolent and shrewd—even by western standards—and was enchanted with their "pleasant wickedness, a sly naughtiness, and an endearingly interesting worldliness." They were a remarkably intact primitive culture, and he planned another visit to make medical, anthropological, and linguistic studies of them.

Meanwhile in England, an American veterinarian, William Hadlow, read Gajdusek and Zigas's report on kuru and noticed that kuru bore a striking resemblance to the disease he was studying. A bizarre affliction of sheep, it had first been reported in Iceland a few years before; it developed slowly, attacked the nervous system, and led invariably to death.

The disease—scrapie—was known to be caused by a virus, but

a virus completely unlike any other. In contrast to the typical short incubation period of other viruses, it was months to years after initial infection that the first symptoms of scrapie appeared.

Hadlow published an article in the journal *Lancet* in which he suggested that in their experiments the kuru researchers had been throwing out the baby with the bathwater. The virus-culture work done on the kuru specimens had been carried out in the traditional way: experimental animals, eggs, and tissue cultures had been inoculated with infective material, and if they failed to show signs of disease within days or weeks, they were discarded. If kuru were a slowly progressing viral infection, as Hadlow thought, then the inoculated animals would have to be kept and observed for much longer periods, perhaps years.

Gajdusek went to Iceland and England to study the methods Hadlow and other investigators were using for the scrapie agent. In goats the incubation period for scrapie was a year and in sheep it was five years. Gajdusek found the prospect of such slow work dismal. He thought it would probably dull the intellect and demoralize anyone who engaged in it. In any case, Gajdusek doubted that a transmissible agent for kuru would be found. There were still too many other possible causes—genetic, autoimmune, metabolic, and toxic—to be ruled out.

Further, as a newcomer to the National Institutes of Health, he needed results in order to assure continued funding for himself. Long-term virus studies did not promise that. All in all, he decided it was an approach he "would prefer to sell to more patient and less ambitious colleagues and collaborators and to await their first successes with skepticism before entering the field." He found a collaborator, Dr. Clarence Joseph Gibbs, a colleague from another NIH laboratory, whom he invited to join his program to help start a long-term research laboratory. Gajdusek, Gibbs, and several technicians planned to inoculate several species of primates and smaller animals with material from the brains of kuru victims and then wait for as long as five years. If nothing happened by then, they didn't know exactly what they would do next. "We decided we would 'reevaluate' the situation and probably kill some of the animals to hunt for virus in their tissues," Dr. Gibbs said.

A patient man, Dr. Gibbs posted a sign in his office at the National Institutes of Health that read: "I'll let you in on a secret—there's nothing really difficult if only you BEGIN."

In 1962 they began—grinding up tissue from the brain of a kuru victim with a mortar and pestle, suspending it in sterile fluid, and injecting it intracerebrally into monkeys. Although chimpanzees were scarce, Dr. Gibbs eventually obtained a supply from Sierra Leone, and in 1963 they were inoculated in the same fashion.

Twenty-one months later, in 1965—eight years after Carleton Gajdusek first saw kuru—the answer came. When Joe Gibbs returned from a scientific meeting in France, he drove immediately out to the Patuxent Wildlife Research Center in Maryland to check on the animals, and there it was: his chimp Daisy had a shaking chill. It was faint, but unmistakable, as if a cloud had passed across the sun and she felt cold. He also noticed that her expression had changed; she seemed withdrawn.

It wasn't long until reports arrived from the animal caretakers that Georgette had started "looking stupid" and that another chimp had dropped the apple she was eating. Daisy and the other inoculated chimps rapidly developed advanced symptoms of kuru, and within months they were dead.

Dr. Gibbs was sorry to see Daisy go. Years later he recalled what a "marvelous little animal" she had been and how she had played around the lab and wrestled with him. But the results were, as Dr. Smadel had predicted, "golden positive." Sadly, he did not live to see them. In the summer of 1963 he died suddenly of cancer.

With the transmission of kuru to chimpanzees, kuru became the first human disease proved to be a slow virus infection. It suggested that many other diseases might also be the work of such unconventional viruses, and evidence for this soon followed.

Gajdusek saw his kuru and unconventional virus work, once so unfashionable, develop into a major field of study.

A number of fatal neurologic diseases found in the West were identified as being of viral origin, and evidence began to accumulate that a vast range of diseases, including forms of dementia and arthritis, might be slow virus infections. In 1968, Gibbs, Gajdusek, and their associates transmitted Creutzfeldt-Jakob disease, a fatal dementia that strikes middle-aged and younger people, to chimpanzees; and viruses have been shown to cause subacute sclerosing panencephalitis (SSPE) and progressive multifocal leukoencephalopathy (PML), both slowly progressive neurologic diseases. Viruses are also suspected, though the evidence is incomplete, of causing Alzheimer's disease—an untreatable deterioration of the mind from which several million Americans suffer and 100,000 die each year—multiple sclerosis, Parkinson's disease, lupus, chronic arthritis, ulcerative colitis, some forms of palsy, Huntington's chorea, amyotrophic lateral sclerosis, Pick's disease, and some syndromes diagnosed as strokes and brain tumors.

Because the kuru agent can be filtered and is self-replicating, it is a virus in the traditional sense. In other respects, however, the viruses responsible for kuru, Creutzfeldt-Jakob disease (which is kuru of the western world), and scrapie are unique. One of their most distinctive traits is that as the virus slowly consumes brain cells, not a trace of response is registered by the victim's immune system. Further, when sections of infected tissue are examined under the electron microscope, no individual viruses can be seen, apparently because of their exceedingly small size. The particles are thought to be about twenty to thirty nanometers, or ten billionths of an inch in diameter, making them the smallest known replicating particles. Unlike other viruses, these unconventional viruses are far more resistant to heat, ultraviolet light, formaldehyde, nucleases, and ionizing radiation than any known human pathogens.

So far, all efforts to purify and further characterize the kuru agent have failed. As the purification process progresses, the virus invariably gets lost. "It's like making whiskey," Dr. Gibbs believes. "It takes an awful lot of sugar and corn to make a little bit of corn whiskey."

But the success of other researchers in purifying extremely small plant viroids gives him hope. Like these viroids—which cause cherry mottling, cucumber pale fruit disease, chrysanthemum stunt disease, Cadang-Cadang disease of coconut palms, and other plant diseases—the kuru agent may be a tiny nucleic acid component, perhaps camouflaged in the plasma membrane of the host cell. This would make it undetectable by the host's immune system. On the other hand, the kuru agent may contain no nucleic acid at all, and consist merely of a bit of membrane that is able to activate preexisting genetic information in the host cell.

One answer revealed with time was why kuru appeared among the Fore, a primitive isolated people in eastern New Guinea. The disease gradually began to disappear in the late 1950s, first among young Fore children and then among the older population. Its disappearance seemed to correspond to a decline in the practice of cannibalism, and the fact that children born after the practice ended did not contract the disease further implicated the cannibalistic rites as the means of spreading kuru.

When Carleton Gajdusek and Vin Zigas first visited the Fore in 1957, the natives had been reluctant to admit they practiced cannibalism. Not only was it outlawed by the government, but the natives themselves had come to suspect the practice. Later work by anthropologists, Dr. Gajdusek, and other researchers revealed that at the time it was practiced more widely than he had first guessed.

As a mourning ritual, the skull of a dead relative was opened and the flesh squeezed into bamboo cylinders and cooked. The work was done by the women and children and almost never by the men or older boys. Even those who did not consume contaminated flesh, which contained more than a million lethal doses of virus per gram, but merely touched the flesh contaminated themselves and others with their unwashed hands. After years of incubation, they became ill.

Many questions remained unanswered twenty years after kuru was discovered. Kuru was eventually transmitted to several species of monkeys, minks, and ferrets, but dozens of other animals and birds never contracted the disease and are apparently resistant. The first rhesus monkey Dr. Gibbs infected in 1962 is still healthy.

Creutzfeldt-Jakob disease is equally puzzling. It is found repeatedly within families, suggesting genetic susceptibility, but environmental factors also seem to be involved: a higher incidence is found among Libyan Jews than European Jews, leading to the conjecture that the Libyans' custom of eating sheep eyes and brain (possibly infected with scrapie) could be a factor. It isn't known either why the Creutzfeldt-Jakob virus is found in most of the body's organs—and even in people not ill with the disease—but causes no apparent harm except in the brain.

As for a cure for either disease, there are no immediate prospects; by the time physical symptoms appear the damage is profound. Vaccines are not possible without a method of early detection, and early detection may not be possible until the particle has been identified, which goes back to purification.

Purification is Joe Gibbs's problem. While he has continued to work in the laboratory, trying to develop better purification procedures, Dr. Gajdusek has concentrated on organizing multidisciplinary teams around the world to investigate many aspects of the slow-virus problem; in addition, he has dedicated time to field work, writing and editing his and his colleagues' papers, and dabbling into distant fields of medicine and anthropology.

For his work on kuru, which revolutionized ideas about what a virus is, Carleton Gajdusek received a Nobel Prize in 1976. By that time kuru had become far more than work to him. The Papuan children stole his heart in 1957, he has said, and his life. His work cost him the woman who would have been his wife.

But as a bachelor he has managed to fill his life with children. He adopted a KuKuKuKu boy and other boys from New Guinea and Micronesia, whom he brought to the United States to be educated.

For many years Gajdusek devoted himself to raising these children. He had a totem pole belonging to one of the children in his living room, and native art hung everywhere in his house. In the winter, the children sat around the dining table studying until late at night; in summer, native music filled the air. He loves children more than anything; with them he is most at ease, and being close to them is his great joy. "As a fat, aging and inept man," he once confessed, "past my youth and my prime, I can still find no difficulties with my Pied Piper tunes, the sincerest notes in my repertoire! . . . All else is but exercise for these tunes, and all work is but practice for the pipes."

6
A Perfect Crime

"THE THING TO DO is figure out the cheapest, simplest, most efficient possible way to test an idea without putting in a lot of time," said David Baltimore. "That's what I did. I had an idea and I thought it was worth a reasonable try. I said, 'I'll give it a try.' I got the stuff.

"If you know what experiments to do, these are trivially *easy* experiments to do. It was not something I had thought about for four hundred years. I went off and tried this one flier. If it hadn't worked, I would have dropped it. . . .

"Of course," David Baltimore added, "I knew full well that if it worked it would be revolutionary."

It worked.

The puzzle was simply this: how does a virus—the smallest and most elemental of life forms—bring about the cancerous transformation of a cell? Sixty years before David Baltimore tried his flier in 1970, a virus was discovered that caused cancer, a deadly sarcoma tumor in chickens. It was a long time before dozens more cancer-causing viruses were found—viruses that produced tumors in rats, monkeys, hamsters, mice, rabbits, frogs, birds, cats, and most animal species except for one. The great cancer-virus hunt failed to turn up viruses that caused the disease in humans.

In a separate line of investigation, molecular biologists set off in the late 1950s to investigate *how* viruses cause cancer. There were two possibilities: either the viruses employed hit-and-run tactics—infecting a cell, tampering with something in the nucleus, and departing—or they behaved like genes and remained in cells generation after generation, issuing their unceasing commands for cancerous cell growth.

Part of the answer came in 1968. Experiments by Renato Dulbecco and two coworkers at the Salk Institute in California showed that certain viruses, like crafty uninvited guests, insinuate themselves into a cell's DNA and proceed to form such strong bonds with the DNA that they become permanent members of the family circle. In this strategic position, the virus's DNA would be reproduced along with the cell's own DNA; whenever the cell divided, the viral nucleic acid would also be passed on to the next generation. If it were a tumor-causing virus, its message of unlimited cancerous growth would be inherited by the new cell. There are viruses, in other words, that masquerade as cellular genes.

The discovery of these viruses, however, was not the complete answer. While it could explain how tumor viruses containing a core of DNA functioned, it could not explain how all the other tumor viruses worked—those containing cores of RNA. Different from cells, which possess both kinds of nucleic acid, a virus contains either DNA or RNA, never both. It is a crucial distinction.

The great obstacle to explaining the RNA tumor viruses was that it is chemically impossible for a virus's RNA to be inserted into a cell's DNA and function as a part of it—DNA must be inserted into DNA, or RNA into RNA. A further difficulty was that according to a fundamental precept of molecular biology, the master planner, DNA, always gave rise to the RNA blueprint, which in turn directed synthesis of proteins following the DNA instructions.

Given these facts, how could RNA viruses ever transform a cell? Yet plainly they did. Most investigators threw up their hands. The question of the RNA tumor viruses was so confounding that few were even interested in it.

A twenty-nine-year-old biologist, Howard Temin, who worked among the sumps and steam pipes in a basement laboratory at the University of Wisconsin Medical School, had a hypothesis about the RNA tumor viruses. He believed that they might be an exception to the rule. While intracellular RNA is not a very stable substance, the RNA tumor viruses acted every bit as stable as the DNA tumor viruses, and they were equally capable of inducing permanent, cancerous transformations in cells, so Temin reasoned that DNA *had* to be involved somewhere along the line. He hypothesized that RNA viruses might be able to transcribe RNA *backward* into DNA, and that this DNA then served as a template for the production of new viral RNA.

It was a farfetched idea, he realized, for it countered everything that was understood about nucleic acid production. The Englishman Francis Crick—who with James Watson elucidated the structure of DNA in 1953 and won the Nobel Prize for this work—had established in brilliant theoretical terms in his Central Dogma that DNA led to RNA which led to protein, the substance of life. A process such as RNA's giving rise to DNA and DNA then giving rise to more RNA was unorthodox.

But Temin thought he saw some evidence for his theory. In cultures he had found that the malignant transformation of normal cells by the chicken sarcoma virus (an RNA virus) could be inhibited by compounds that inhibited the manufacture of DNA. It was flimsy evidence at best; nevertheless, Temin proposed his idea before a gathering of virologists at Duke University in 1964.

His talk was not well received. Although Howard Temin was considered one of the brightest young scientists, it was the general consensus that this time he was barking up the wrong tree. Temin made matters worse by zealously insisting on his theory in conversations over breakfast, lunch, and dinner during the meetings and, despite the fact that he was pointedly ignored, in streams of subsequent reports and lectures.

A former student of Temin's, David Baltimore, had listened with interest to one of these talks, although Baltimore did not work with tumor viruses. Like most, he regarded the question of the RNA tumor viruses as much too difficult—if one cared about results.

Baltimore was a biochemist, and he liked enzymes. Working at the Rockefeller Institute he had helped demonstrate that certain viruses could induce synthesis of an enzyme that did something no cell could accomplish alone: manufacture RNA from RNA. The virus Baltimore studied was a close relative of poliovirus, and he went on to study poliovirus itself.

A postdoctoral student who came to work with him tried to interest Baltimore in another virus, vesicular stomatitis virus (VSV), an odd, bullet-shaped creature that produces a foot-and-mouth ailment in animals. The student, Alice Huang, a Chinese-born graduate of Johns Hopkins University, was a virologist, and she had come to Baltimore's laboratory at the Salk Institute in California to study his biochemical techniques. Having written her Ph.D. thesis on VSV, she wanted to continue working with it.

"I said no," Baltimore recalled. "I said I didn't know anything about VSV. I told her, 'Work on poliovirus, which I know something about, while you are here, so at least you learn what you came to learn.' "

They worked on poliovirus, and Baltimore and another student in the laboratory proceeded to discover some of its remarkable secrets. They found that the entire viral genome—all of its genes—was not translated bit by bit into proteins but was instead translated into one huge continuous polyprotein, which the virus then chops into small handy pieces. Baltimore believed that if the reason for this could be understood, a major clue to the virus might be revealed.

Although her virus did not captivate him, Alice Huang did. When Baltimore subsequently accepted a post as associate professor at the Massachusetts Institute of Technology in Cambridge, a position was also arranged for her, and they were married.

At MIT they decided that Alice would apply the successful poliovirus techniques David had developed to the study of VSV. This virus was a curiosity: it produced defective viral progeny that inhibited further viral growth.

At first the studies were done by Alice Huang and a graduate student, Martha Stampfer. "Martha was remarkable," Baltimore said. "She found that VSV had surprises in it we didn't even guess were there." She learned that VSV manufactured nine different types of RNA, an unheard of number at that time. The researchers did not quite believe it themselves, and in reporting on the work they said it seemed exorbitant. But not until another laboratory reported that the nucleic acid of VSV consisted of a negative strand of RNA—a template, which is a meaningless piece of information in itself—did Baltimore suddenly become interested in the virus.

"I began thinking about the puzzle," he said, springing up to draw on a brown chalkboard in his tidy modern office at the Massachusetts Institute of Technology. "RNA can either be positive or negative, either the same or the opposite polarity to the messenger RNA generated from it."

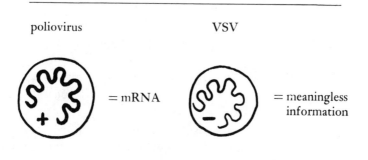

poliovirus   VSV

"With poliovirus, the virus *is* immediately messenger RNA. But with VSV, the opposite is true." He asked himself how viruses like VSV, that carried only the mirror-image of messenger RNA, could ever start an infection. The negative strand of VSV, being a mirror image, is meaningless information. He drew:

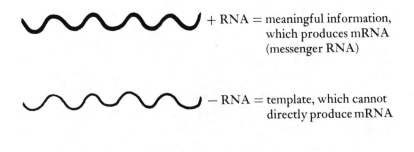

$+$ RNA $=$ meaningful information, which produces mRNA (messenger RNA)

$-$ RNA $=$ template, which cannot directly produce mRNA

"There seemed to be two possibilities," he went on. "The RNA of VSV could enter the cell and be copied by an enzyme provided by the cell to make messenger RNA, *or* the RNA could come into the cell carrying such an enzyme with it. There was, and is, no evidence that normal cells have ever transcribed RNA from RNA, so the most interesting notion was that the enzyme was *in* the virus."

The idea that the virus might bring along an enzyme when it invades a cell was not unprecedented, but the two viruses that seemed to do so were suspected of being quirks, experimental oddities.

"The idea that the enzyme was *in* the virus got me interested," Baltimore said, "because I knew how to test it. I said to Alice, 'Let's make a big prep, and we'll test it.' "

They grew the virus in cells from a Chinese hamster ovary and purified the viral particles. Then they poured detergent into the flasks of purified viruses, stripping away the protective viral envelopes and exposing the cores. Next, the naked viruses were incubated with radioactive GTP (guanosine triphosphate), one of the precursors of RNA, at body temperature for twenty minutes. If the viruses themselves contained an enzyme that could make RNA, the radioactive GTP would be used by the virus and incorporated into the RNA. The virus mixture was then chilled in an ice bath to stop the reaction. Acid was added, precipitating the RNA, and the amount of radioactivity in the newly made RNA was measured. Baltimore found that the radioactive GTP was in fact incorporated into RNA and that the virus was necessary to drive the reaction.

The experiment was completed by David Baltimore, Alice Huang, and Martha Stampfer in a couple of hours. Testing more viruses, they found evidence that they, too, contained an enzyme that made RNA. Their investigation led to the realization that a large family of such viruses exists—negative-strand viruses that import into cells an enzyme enabling them to transcribe messenger RNA from RNA.

Then an exciting thought occurred to Baltimore: why not look for a virus-imported enzyme in the RNA tumor viruses? If one could be found, it might help explain how the viruses wrought their malignant transformations.

He recalled Howard Temin's theory: "RNA to DNA to messenger RNA? It seemed baroque, and it countered the accepted dogma. But although no one had ever *seen* reverse transcription as a biologic process, there was nothing outlandish biochemically about it. Nothing outlandish at all. The outlandish was more in saying it. It *could* happen. It was a reasonable notion."

But "Teminism," as the theory had been branded, did not have many supporters; and Temin had become something of a zealot about the notion.

Still, Baltimore had admired Howard Temin ever since first meeting him. As a high school student, Baltimore spent a summer at the Jackson Laboratory in Maine, where the slightly older Temin taught; and when it came time to choose a college, one reason Baltimore had selected Swarthmore over Harvard was because Temin had attended Swarthmore. The other reason was that he did not like the "rah-rah attitude" in the freshman dormitories at Harvard. At Swarthmore he sensed "a more obvious commitment to the intellectual life, a seriousness about intellectual activities, about life, that was good."

Baltimore's mother understood the choice, but his father, a manufacturer of women's coats, did not at first. Both parents placed great value on education, and at considerable inconvenience they had moved out of New York City to a Long Island community where the schools were thought to be superior when David and his brother were young. Later, his mother had returned to school herself to train as a clinical psychologist, and it was she who had heard of the Jackson Laboratory program and sent Baltimore there one summer "to keep me busy."

"It was a typical Eastern European Jewish background," Baltimore responded in his succinct, very efficient manner when asked once about his family background. "And that background is unthinkingly encouraging of intellectual activity."

Baltimore had no particular reason to think Temin was wrong. Unencumbered himself by any experience in the field of tumor virology, he had no stake in one position or another, and thus no ax to grind. If there were an enzyme in the VSV particles, it seemed reasonable to Baltimore that there might be an enzyme in cancer viruses.

But those who were most knowledgeable about tumor viruses believed that the viruses carried only the coding potential for enzymes, not the actual enzymes. In the case of poliovirus, Baltimore had found this to be true. Nonetheless, he made up his mind that if he could obtain some tumor viruses he would run a few quick experiments—nothing that would take much time.

He also decided to keep his options open: he would search for either RNA *or* DNA enzymes—polymerases—in the tumor viruses. It was more likely that the tumor viruses were making RNA from RNA, so Baltimore hunted first for RNA polymerase in a chicken sarcoma virus. But he detected no enzyme activity.

After being given some mouse leukemia virus by the National Cancer Institute, he tested again, this time for DNA polymerase —an enzyme, contained in the virus, that could transcribe DNA from RNA.

It took no time at all to find what he was looking for. "It took a little longer to tie up the pieces," he said, "but it was very clear that it was correct in a week."

The revolutionary experiment resembled the VSV experiment, but instead of using radioactively labeled GTP, Baltimore used TTP (thymidine triphosphate), one of the precursors of DNA. Purified mouse-leukemia virus from the plasma of infected Swiss mice was incubated at body temperature with the TTP. After forty-five minutes the mixture was chilled, the DNA was precipitated out with acid, and the amount of radioactive TTP made into DNA was measured: there was much more of it than in an unincubated reaction mixture.

Baltimore repeated the experiment, which took about an hour, several times a day for several weeks, and he ran dozens of control experiments. All the results pointed to one explanation: a unique enzyme, unknown in any cell, was catalyzing the synthesis of DNA from RNA.

This extraordinary piece of machinery owned by the tumor virus itself—reverse transcriptase—explained how RNA viruses could bring about the malignant transformation of cells. Once the viral RNA was converted to DNA, it would function exactly as Renato Dulbecco had described—by becoming a provirus, integrating itself into cellular DNA and behaving thereafter like a gene.

Baltimore did not waste any time when it came to writing up a report on his discovery of reverse transcriptase. But before mailing it off to the British journal *Nature*, he remembered the one person he ought to tell about it: Howard Temin. For years

Temin had been trying to prove his own theory, and he had tried many different approaches, none of which had worked.

By the rarest of coincidences, Baltimore learned when he called him, Temin had also just completed experiments that supported his hypothesis. Temin did not have all the necessary results, but he was well on his way.

Baltimore went ahead and submitted his paper to *Nature*, and publication was delayed until Temin's paper arrived. Such is the premium placed on being the first to make a scientific discovery that the date a paper is received by the editors appears when it is published. Baltimore's report, received June 2, 1970, and the paper of Temin and his associate, Satoshi Mizutani, which was received ten days later, on June 12, 1970, appeared together on June 27, 1970.

Even with the simultaneous publication of two independent reports, the editors cautioned that the findings were "preliminary" and "heretical," suggesting that the classic principle of information transfer from DNA to RNA was an oversimplification that had delayed progress in the field for more than a decade. At the same time, *Nature*'s editors predicted that the discovery, if upheld, was likely to cause an uproar in molecular biology. The editors further predicted that if reverse transcriptase proved unique to tumor viruses, an important new kind of cancer chemotherapy could emerge, based on compounds that would inhibit the enzyme's activity.

Within hours of the journal's arrival in laboratories around the world, reverse transcriptase experiments were under way. One of the first confirmations of the findings came from an authority on tumor viruses at Columbia University, Sol Spiegelman, who had taken a wrong tack on an RNA-dependent RNA polymerase.

"After we published," Baltimore recalled, amused ten years later at the blaze of interest the little enzyme provoked, "Spiegelman came in right behind us. Within *days*, he had confirmed our findings, and a lot of other people quickly moved in and adopted

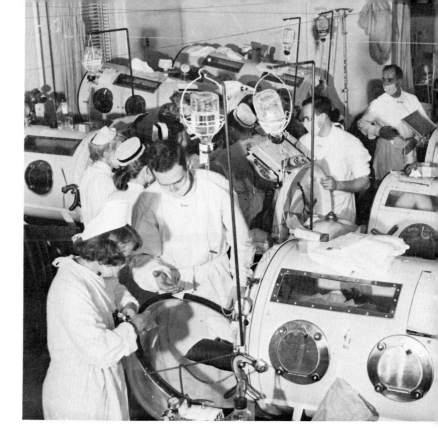

August 1955. An emergency polio ward at Haynes
Memorial Hospital, Boston, during an epidemic.
(*March of Dimes Birth Defects Foundation*)

Jonas Salk, who developed the first polio vaccine, at
the University of Pittsburgh. His vaccine was
administered nationwide in the spring of 1955.
(*National Library of Medicine*)

Sir Christopher Andrewes, 1950. "I felt that the time was ripe for having another try at cracking the nut" of the common cold. (*Ronald Procter*)

Carleton Gajdusek with friends in New Guinea, 1957.
(*National Institutes of Health*)

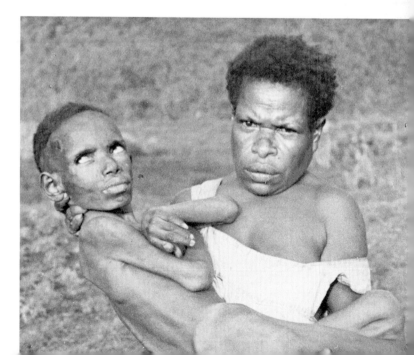

"The thing to do is figure out the cheapest, simplest, most efficient
possible way to test an idea without putting in a lot of time.
That's what I did."–David Baltimore at MIT in 1975, the year
he won the Nobel Prize. (*Calvin Campbell: Massachusetts
Institute of Technology*)

A victim of kuru, about eight years old, seen by Carleton Gajdusek
n New Guinea in 1957. "The natives have given up on our medi-
ine," he admitted after months of trying to cure the strange
ffliction. "They have decided that this magic is too strong. . . ."
National Institutes of Health)

For years Howard Temin had been trying to prove his theory
about certain cancer viruses, which was widely dismissed as
"Teminism." By rare coincidence, he and David Baltimore reached
the same conclusions almost simultaneously. Howard Temin at the
University of Wisconsin at about the time he won the Nobel Prize.
(*A. Craig Benson: National Library of Medicine*)

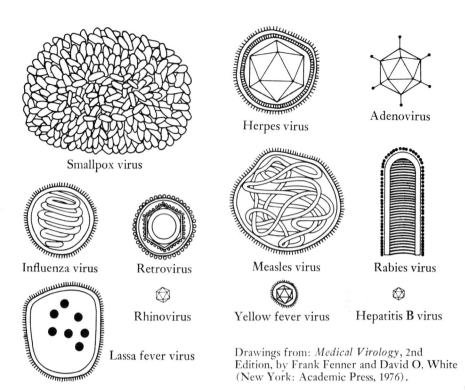

Smallpox virus

Herpes virus

Adenovirus

Influenza virus

Retrovirus

Measles virus

Rabies virus

Rhinovirus

Yellow fever virus

Hepatitis B virus

Lassa fever virus

Drawings from: *Medical Virology*, 2nd Edition, by Frank Fenner and David O. White (New York: Academic Press, 1976).

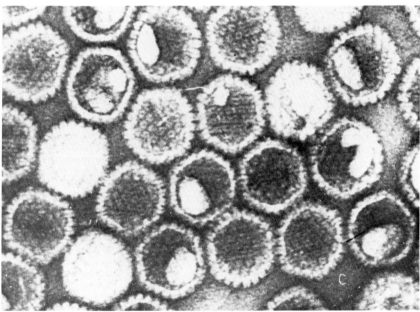

Herpes simplex viruses magnified more than 20,000 times by electron microscope. (*Center for Disease Control*)

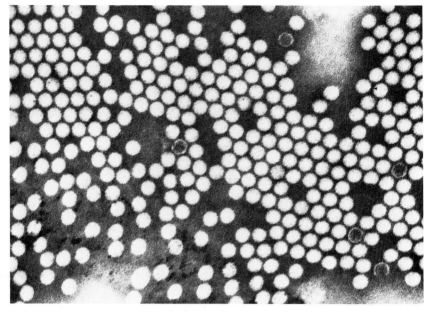

Polioviruses. (*Center for Disease Control*)

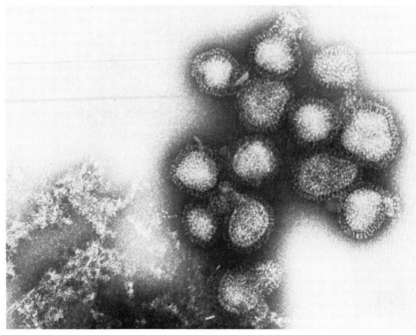

Influenza viruses with their surface spikes visible. (*Center for Disease Control*)

our approach. Almost immediately, reverse transcriptase became a national industry."

Shortly thereafter, Spiegelman and his group found evidence of reverse transcription in six different RNA tumor viruses. The discovery stirred great excitement. One of the major disappointments of virology is that so often there turns out to be little if any correlation between a virus's physical characteristics and the disease it causes. But with the RNA tumor viruses it became clear that for once there was a stunning correspondence between form and function: all of the RNA viruses capable of malignantly transforming cells possessed reverse transcriptase!

These viruses which transcribe DNA backward—the retroviruses—have been shown to cause cancers in chickens, cats, cows, mice, gibbon apes, and numerous other animals and birds. Like efficient businessmen, as Baltimore pictures them, the retroviruses enter a cell carrying along the tools of their trade—thirty to fifty molecules of reverse transcriptase. Using the virus's genome, a single positive strand of RNA as a template, the enzyme manufactures a complementary strand of negative DNA. This negative strand in turn gives rise to a complementary strand of positive DNA, and there it is—double-stranded DNA looking exactly like cellular DNA. The viral DNA then journeys from the cytoplasm into the cell nucleus and inconspicuously inserts itself into a cell chromosome.

Now, as a provirus the viral DNA produces RNA, which, like any RNA, migrates to the cell's cytoplasm where, as messenger RNA, some of it directs synthesis of viral proteins; the rest forms the genome of new viral particles. Migrating to the edge of the cell, the viral RNA and proteins wrap themselves up in some of the cell membrane and break away. Cunningly camouflaged in their cell-membrane coats, new young viruses enter the world. The entire extraordinary performance takes only hours.

For elucidating how viruses can cause cancer, David Baltimore, Howard Temin, and Renato Dulbecco shared the Nobel Prize in

1975. The joint discovery of reverse transcription by Baltimore and Temin was an unusual example of friendly competition among scientists. Howard Temin was delighted that his idea had been exonerated; he had felt frustrated at the failure of his colleagues to appreciate the idea for so long. "However," he said some years later, "I now understand this and realize that it is part of the dynamics of science to resist unorthodox ideas, since most of them are incorrect."

David Baltimore was also delighted. "There was no rivalry between Howard and me that I know of," he said. "We were complementary. It was his concept and my enzymological knowledge that proved it. Howard had been trying for years to prove it. When he adopted a style that I had set in biology, then he had success." For his part Temin says graciously that Baltimore "certainly deserves any credit that he gets."

For Baltimore, as for Temin, there was much gratification in the reverse transcriptase work. "I am told that our paper on reverse transcription is one of the ten most often cited scientific papers to be published in the last ten years," Baltimore says, "and since ninety percent of all scientists who ever lived are alive today, that means it is one of the most cited papers in the history of medicine. Although—" he adds, reining himself in, "I don't know, Einstein . . . that may not be fair."

The discovery brought him to the forefront of cancer research at a time when cancer was one of the few magic words still able to open the big funding purses. He was promoted to full professor at MIT at the age of thirty-four, and he was able to vastly expand his laboratory, eventually to a staff of twenty, with experts in many specialties, including recombinant DNA technology and immunology.

"There is nothing like success to increase one's interest in a subject," Baltimore says amiably. However, he entered the field and expanded his operations well aware that in doing so he would have to spend all of his time administering his empire while the actual laboratory work would be done by others. He felt he had no choice but to expand. "The only way to get into RNA tumor viruses was to get into it in a big way. Those who have been successful almost always have big labs."

Following the discovery of reverse transcriptase, David and Alice took a month's vacation in the Orient and Nepal—his longest vacation ever. Although he savors working in a highly competitive, fast-moving field where there is continual discourse and a sense of progress, he admits that "the other side of it is that you can't relax very much"—he slices the air with his hand—"or the whole field will go 'bye."

Sheltered within cellular DNA, retroviruses lead a covert existence. They are certainly in a position to make permanent alterations in the genetic character of a cell. But *do* they? Baltimore's faith in the retroviruses as agents of carcinogenesis is based on their "exquisite self-control." It would seem that the DNA viruses should be equally able to transform cells; and there are DNA viruses that under certain experimental conditions do so. But under natural conditions they seldom do.

The explanation, Baltimore believes, is that unlike his pet RNA viruses, "the DNA viruses are greedy." In the process of multiplying they usually kill the host cell, and where they succeed in bringing about transformation they seem to lose the knack of reproducing themselves. Only herpes viruses, apparently, can both call the tune and play the music—and are therefore a leading suspect among the DNA viruses.

The retroviruses are not greedy; they are usually quiet, inconspicuous guests, and their reproductive rites do not usually exhaust the host cell, but instead usurp only a tiny fraction (about one percent) of its resources. The retroviruses appear to direct synthesis of a "transforming protein," but how this protein functions, where in the cell it works, and why a tumor virus codes for such a seemingly unnecessary substance are mysteries, for strains lacking transforming protein appear to survive equally well.

In mice, retroviruses evidently infiltrate the germ cells—sperm and ova—and can be passed along to offspring as silent, unexpressed genes, suggesting that the retroviruses do become true genes. Whatever the role of viruses in the transformation of cells, other factors are involved. In some studies of mouse leukemia virus, the provirus was not expressed except when stimulated by chemicals. Other viruses, such as mouse mammary tumor viruses, seem sensitive only to hormones, and once stimulated they seemingly can transform only specific cells, the mammary cells. Other investigations indicate that the proviruses may be incited to action through the combining of a newly infecting virus with some pre-existing proviral DNA.

Inherited proviruses appear to be very widespread—as many as a hundred different proviruses and other bits of viral genetic information have been found in normal animal cells. So it seems that proviruses play a role in transforming cells only rarely—perhaps in certain kinds of cells, in the presence of a second virus or chemicals, with certain cellular immune responses operating, or most likely, with some combination of these factors.

What, then, are the proviruses, scattered here, there, and everywhere, up to the rest of the time? Howard Temin believes they form some vital part of an animal's genetic complement, occasionally turning into retroviruses, but otherwise playing an important hidden role in a cell's biologic processes. Others hold that proviruses evolved from an ancient cellular infection, and having taken on some important role, they remain in cells as a result of natural selection. David Baltimore maintains that viral genes could have been inherited for millions of years without offering any selective advantage; all that would be required is that their negative effect be slight—which it is.

But this is speculation, and it has not brought the important answers any closer. So far, no retroviruses or any other viruses have been proved to cause cancer in humans. "Had cancer in humans only turned out to be like animal cancer," Baltimore laments, "by now we could have made a vaccine against it."

The circumstantial evidence is compelling, as compelling as

the fingerprints on a murder weapon, but as in the case of the herpes virus and cervical cancer, it stops short of closing the argument.

It was noted by an Italian physician as long ago as 1842 that nuns rarely suffered from cervical cancer and that the disease was much more common in married women than unmarried ones. More recently, women who have had more than one sexual partner were found to be more susceptible to the disease—a clue that cast suspicion on a transmissible agent.

Then in the 1960s studies of mothers with babies who were born with herpes simplex infections led researchers to make a connection between genital herpes infections and cancer. Members of the herpes (in Greek, "to creep") virus family cause a variety of illnesses from small sores around the mouth, to genital infections, infections of newborns, central nervous system infections, and blindness. They may also cause some abortions, birth defects, chronic neurologic diseases, and perhaps some cancers.

In cervical cancer, the virus has not been caught in the act, but its shadows and footprints are everywhere. In cervical cancer patients, antibodies to certain herpes simplex viruses (HSV-2) are found much more often than in normal women; and when the cervical cells begin to deteriorate, they yield HSV-2 particles.

The most convincing link so far centers on an exotic tumor discovered by an Irish surgeon who worked in Uganda in the 1950s. Dr. Denis Burkitt saw numerous young African children, usually about eight years old, whose jaws were swollen as if they had mumps. It was a malignant tumor that occurred, he learned, only among inhabitants of warm, wet, low-lying areas.

Because of its distribution and the absence of other signs of infection, Dr. Burkitt proposed that the tumor was caused by a virus and transmitted by a mosquito. Several years thereafter, two researchers in London isolated a virus from one of Burkitt's patients. To their surprise, it bore no resemblance to other insect-carried viruses: it was a herpes virus. Named EB virus for its discoverers, Dr. M. A. Epstein and Dr. Yvonne Barr, it has been isolated consistently from Burkitt's lymphoma patients ever since. Tumor cells from Burkitt's patients nearly always contain the EBV genome and some EBV antigens; and high levels of antibody to the virus have been found in patients' blood.

While all of this mightily incriminates the virus, there are other pieces of evidence that do not quite fit. For one, the EB virus has never been found to be carried by a mosquito. For another, the virus has been detected in normal healthy people around the world; yet it is associated with cancer only in inhabitants of parts of Africa and Papua New Guinea. Further confusion comes from the association of this same EB virus with cancer of the nose and throat in Southern China and with mononucleosis; patients with these diseases have unusually high levels of antibody to the EB virus.

Herpes simplex viruses do cause cancer in some animals, and if either herpes simplex or the Epstein-Barr virus could be found to satisfy Koch's postulates—the classic criteria to *prove* an infectious agent as the cause of a disease—then the first human cancer virus would be found. The virus would have to be isolated from a human tumor, cultured, purified, and injected into another person and then be found to produce exactly the same type of tumor. The virus would then have to be isolated from the new tumor and shown to be the same as the original virus. These are the criteria that must be satisfied. Short of this, the circumstantial evidence must be as overwhelming as it is for tobacco smoke in lung cancer.

With deceptive signals and roadblocks placed by nature in every direction, fuller understanding of the intimate relationship between cells and viruses may be very close, or a long way off. This does not greatly concern researchers like David Baltimore, because extraordinary revelations come often at the molecular level. When reverse transcriptase was discovered, only evidence of its activity was seen; it was not known what it is or how it works. "We now understand the mechanics of reverse transcrip-

tion in *detail!*" Baltimore says excitedly, describing the fruit of
ten years' work.

"We also know that the transformation of cells is a highly
selective process, with each type of tumor virus only able to
transform a small range of cell types. So it must be that each
type of transforming virus makes a very specific type of transforming protein. . . . We also have the life cycle of the retroviruses in fair detail. . . ." As he speaks, he snaps a rubber band
until it breaks and flies at his visitor, but he deftly reaches out
and retrieves it in midflight. "And a normal cell gene has been
found that is incorporated into the tumor virus. The virus ac-
quires a new genetic capability from a normal cell! What does
that mean? Why this should be is a fascinating question!"

Baltimore would love to know the answer. He would also love
to solve the puzzle of how the transforming protein works in a
strain of mouse virus that turns a perfectly normal cell into a
leukemia cell. "Someday," he says, "I'd like to walk into the
laboratory and do the reverse transcriptase experiment that will
answer this question." He is confident that the answer will come
in the next five years, though not at all confident that he will be
the one to find it.

The uncertainty of the work and the snail's pace of progress
toward medically useful discoveries do not daunt him in the least.
As a virologist, he counts himself among the luckiest of scientists
because he can come to know his "chosen pet" down to its very
molecules, and he marvels at its supreme efficiency, economy,
and elegance.

"You can't be successful in this work," he offers, jumping up
from his chair to indicate that our visit is over, "unless you begin
to love the viruses."

7
Pay Dirt

ONE LATE FALL DAY in 1960, Cicillo Aldo took a plastic spoon out of his pocket, bent down, and scooped up a spoonful of soil near the town of Boscotrecase, a dozen miles southeast of Naples, Italy, in the shadow of Mount Vesuvius.

As he had been instructed to do, Cicillo Aldo put the soil into a plastic bag, sealed it, placed the bag in an envelope, and sent it off to Detroit, Michigan. There, in one of a dozen old brick buildings barricaded behind a tall iron gate at the edge of the Detroit River, the arrival of his packet was duly noted before it was dropped into a cardboard box and filed away on a closet shelf along with hundreds of others that looked just like it.

Cicillo Aldo, a salesman for the Milan branch of Parke, Davis & Company, had forwarded the sample to the pharmaceutical firm's Detroit headquarters as part of their soil collection and screening program. In the search for medically useful microorganisms, it was the company's policy to leave few stones unturned, and samples were gathered at random, anywhere and everywhere by company employees. The program had been initiated in 1950, and with hundreds of ambitious Parke, Davis salesmen traveling

the world armed with spoons and plastic bags, the company accumulated soil samples by the tens of thousands.

Researchers, laboratory technicians, and executives joined in,
and one enthusiastic director of biological research returned from
a trip to South America with ninety specimens. Two organic
chemists went deep into the Florida Everglades in quest of exotic
microorganisms. "We thought we'd get something very different," said one of the chemists afterward. "The Everglades trip
didn't amount to a hill of beans."

The method was, despite its apparent randomness, highly
rational—and worthwhile. The earth teems with microorganisms,
and some of them happen to be exceedingly useful: they manufacture antibiotics—chemicals that destroy or otherwise interfere
with the growth of other microorganisms. Penicillin is an example of such antibiotic growth; it is manufactured by the fun-
gus penicillium.

Penicillin had been discovered by complete chance, in 1928,
when a common mold contaminated the bacterial cultures of the
English researcher, Dr. Alexander Fleming, and interfered with
their growth. The discovery, in similar fashion, of other early
antibiotics had set off the hunt in the pharmaceutical business for
more of the beneficent microorganisms. Chloramphenicol, the
first important antibiotic developed by Parke, Davis, was found
in a random soil sample, and over the years the firm had devised
an elaborate, painstaking system of screening the samples. Most
often, the organisms unearthed were familiar bugs already in use
and patented by a competitor. But the more samples that were
examined, the better were the chances of finding pay dirt.

For two years, while the slow methodical sifting of soil samples
proceeded in Detroit, Cicillo Aldo's sample remained in a box. In
turn, it was eventually taken out and examined. For this process,
a gram of the dirt is put in a test tube with distilled water and
centrifuged. Tiny drops of the muddy water are then placed in
petri dishes containing agar, a growth medium, and after several

days of incubation a bright patchwork of microbial colonies springs up. A sample from each colony is examined under a microscope. If more than one kind of microorganism appears in a colony, the process is repeated until pure cultures are obtained. In the culturing process antibiotics are used to inhibit the growth of contaminating fungi and other undesirable organisms.

Once a pure culture is in hand, the next step is to try to make more of it. The organism is inoculated into a small flask containing a nutritious fermentation medium of sugar, nitrogen, and minerals. The flask is covered with gauze and gently shaken by machine for about five days to encourage the organism to grow. The contents of the "shake-flask" are then centrifuged to separate the organism from whatever it might have made, if anything, and the clarified fluid is tested against a microbial screen—more petri dishes, seeded with bacteria—to determine whether it is active against any of them.

In the petri dishes the formation of a clear area around the drops of the fluid indicates that antibacterial properties are present. When this occurs, the experiment is repeated for confirmation. The next stage is to eliminate all known antibacterial agents (by the early 1960s there were about 2,500 of them) through further screening. At this point nearly all the soil specimens are discarded.

Cicillo Aldo's sample was not thrown out, but because the organism had not multiplied well in the flasks and its antibacterial activity was so weak, attempts to identify the organism had failed. Once they learned how to concentrate it, researchers were able to identify Cicillo Aldo's specimen: it was a strain of the bacterium *Streptomyces antibioticus*. Members of this family manufacture a number of important antibiotics, among them streptomycin and chloramphenicol. Preliminary studies of the active component in the new strain showed a resemblance to chloramphenicol. But it also seemed to have antibacterial properties that chloramphenicol did not have.

"Potentially novel" was scribbled on a flask of the substance, and it was sent on to the chemists. Their job was to concentrate it further, purify and crystallize it, and determine exactly what the agent was. Just before Christmas 1962 the first large quantity of the agent was harvested from a stainless steel "stirred-jar," a fermentation container stirred with motorized paddles. The agent was designated J-87, and analysis began.

The chemical analysis of "cuts," or small samplings, from the jars is slow work, and the difficulties with J-87 were numerous. It was two long years before pure crystals of J-87 were obtained, in March of 1964.

But it was a dismal moment. Its frail antibacterial properties had vanished! And despite repeated efforts they couldn't be recaptured.

Rather than lamenting the loss, someone in the chemistry division thought to pass J-87 on for a last chance to the company's new antiviral section. Was it curiosity, good sense, chance or just a matter of routine that it was sent? No one can recall. "Success," said Henry Dion, an organic chemist who directs antibiotic research at Parke, Davis and who later worked on J-87, "was a combination of everything."

In July 1964 the antiviral section issued a surprising report on J-87. "*Lo and behold!*" Henry Dion later exclaimed, remembering the moment. "It showed *tremendous* activity in the antiviral screen! *But,*" he hastily amended, "if you're in this racket, you don't trust the first sample of anything. You send a second."

The second antiviral assay confirmed it. J-87 was unmistakably active against a virus: herpes simplex. But, it was all too well known that although virus-killing activity in cell culture was not very hard to find, it invariably failed to be confirmed in living beings. Ordinary vinegar would kill isolated herpes viruses, but once the viruses penetrated the cells of a living animal, the only way to get at them was to kill the cells. An antibiotic such as penicillin prevents the synthesis of bacterial cell walls by interfering with the formation of links between the wall's units; without its rigid coat, the inner bacterial cell membrane pokes through and the organism explodes. But an agent that could differentiate between a virus sequestered inside a cell and the cell itself was judged to be an impossibility.

Even if one existed, there would be other impediments to its success. One is the speed with which most viral infections set in: by the time symptoms appear, the infection is near its peak, and by the time a diagnosis can be made and treatment started, the infection is already waning and the patient's fate has been determined. Further, the same symptoms can be produced by many different viruses, and in two people the same virus may produce different symptoms, making diagnosis of the precise virus difficult.

For all these reasons it was acknowledged that viruses could not be fought with drugs and that it was pure fantasy to think an antiviral agent could be found that would be effective against a broad spectrum of the microorganisms. The director of clinical research at Parke, Davis, Dr. Robert Buchanan, believed this. He had learned it in medical school and nothing in his years of experience as a pediatrician had shown him differently: "No drug can cure a viral infection."

Cancer researchers had developed certain nucleosides—they are the building blocks of nucleic acid—that they used to try to inhibit the growth of tumor cells, and it seemed to some that they might work against viruses. But the first efforts along these lines only fortified the argument that a nontoxic antiviral agent was a fundamental contradiction in terms.

This argument was challenged in 1962 when a Boston researcher, Dr. H. E. Kaufman, cured a blinding herpes eye infection with a nucleoside, a derivative of uridine. It was the first time in history that a drug cured a viral disease. The compound, idoxuridine, did not work in every case, however, and it had other serious drawbacks. Most significantly, it could not be given internally because of its extreme toxicity to healthy cells; so its use was very limited. Nevertheless, the work with idoxuridine led researchers in laboratories around the world, including Parke, Davis and other drug companies, to initiate antiviral research.

A few months after the promising report on J-87 at Parke, Davis in Detroit, two French researchers reported discovering an agent, a nucleoside, that could suppress viruses.

The story of this nucleoside had begun years before, with a small black marine sponge. The sponge, *Cryptotethia crypta*, had been found in 1945 by a chemist, Werner Bergmann, who had been gathering sponges in shallow water off the Florida Keys. Not recognizing the specimen, he collected several and preserved them in a solution of formalin and seawater.

Back in Dr. Bergmann's laboratory at Yale University, the sponges were dried in an oven. More were found three years later, and they were identified as a new species. Two years after that, *Cryptotethia crypta*'s secret was extracted by Bergmann and his assistant. After melting, stirring, boiling, cooling, crystallizing, recrystallizing, benzoylating, acetylating, debrominating, saponificating, hydrolyzing, hydrogenating, evaporating, refluxing, distilling, diazotinizing, shaking, washing, and titrating the sponges, they were left with "a nice crystalline material" that contained two brand-new compounds.

They were arabinosyl nucleosides—derivatives of thymine and uracil, which are two of the bases of nucleic acid. These nucleosides had a fascinating property: they were capable of subverting cellular metabolism, specifically, the production of nucleic acid. It was widely appreciated that if exactly the right kind of nucleoside were found or made, it might be used to kill cancer cells.

Many laboratories began trying to synthesize such a nucleoside. Nucleosides consist of bases—adenine, guanine, cytosine, and thymine or uracil—bound to sugar molecules. In RNA the sugar is ribose, in DNA it is deoxyribose. In the arabinosyl nucleosides it is the uncommon sugar arabinose. It was hypothesized that when DNA—two sugar–phosphate chains linked by its bases, like a ladder by rungs, and the whole twisted in helical form—was synthesized in a cell, the fake sugar, arabinose, would play havoc. Something like baking soda mistaken by the cook for baking powder, with the result that the cake does not rise, sugar and phosphate would not bond properly, and DNA would not form. The cancer cells would not multiply.

One of the first nucleosides to be developed was cytosine arabinoside, and it showed great promise in fighting leukemia. Another new nucleoside, adenine arabinoside, was first synthesized at the Stanford Research Institute in California in 1960 (the same year Cicillo Aldo scooped up his dirt sample). Tests of this nucleoside initially indicated that it halted the growth of tumors in mice. But further studies soundly disproved this, and the California cancer researchers discarded adenine arabinoside.

Research on adenine arabinoside was picked up in France at the Centre de Recherches des Laboratoires Diamant, a private pharmaceutical firm. There, Jean de Rudder and Michel Privat de Garilhe tested adenine arabinoside against a standard antiviral screen. They reported its remarkable effects in October 1964: the nucleoside completely suppressed two viruses, herpes and vaccinia (a close relative of smallpox virus that was used to make smallpox vaccine)—even massive quantities of the viruses. In concluding their report, De Rudder and De Garilhe marveled at the complexity of the agent involved. But they were not heard from again; presumably, they had not realized the significance of what they stumbled on.

At this point, no one at Parke, Davis knew what J-87 was, but with the arrival of the French report in the fall of 1964, the J-87 project moved into high gear. The microbiologists began studying ways of making bigger batches of J-87 by altering the ingredients, temperature, and dozens of other factors in the fermentation process. The following spring the first 2,000-gallon tank of J-87 "beer" was harvested in Detroit.

Chemical analysis of the beer was completed several months later. To the distress of everyone who had worked on the project, J-87 was none other than adenine arabinoside! As a known compound which had first been synthesized in California, Parke, Davis would not be able to patent it.

Work on the project proceeded anyway, according to a company executive, partly because once the process got going and all the various divisions (chemistry, fermentation, tissue culture, and so on) got busy with a project, it was almost impossible to stop it, and partly, too, because of the agent's unusual antiviral powers.

Further testing confirmed that J-87 killed herpes as well as vaccinia viruses. The next logical step was animal studies.

In mice, as in the test tube, J-87 combatted herpes and vaccinia infections with ease. Rats, rabbits, guinea pigs, and hamsters were tested next, and the compound was applied to animals' eye infections. J-87 continued to look good, and a battery of more challenging studies was designed. J-87 was given orally, intravenously, and intracerebrally to many different animals as well as spread on their skin. In rabbits, it cured severe vaccinia infections; in mice, it combatted herpes infections of the central nervous system and brain. Animals who were given lethal doses of herpes viruses were saved with injections of J-87. And, as far as anyone could tell, there was no sign of toxicity to healthy cells.

There was one great, obvious drawback: even in the tiniest of animals, the dose necessary to cure an infection was enormous. "In humans," Henry Dion said, "it looked as if you would have to use a shovelful."

There were other suspicions about the compound, but the consultant who had directed the hundreds of animal studies for the company was excited. "You've got a heck of a successful antiviral agent here," he kept saying. "Let's move, move, move!"

Metabolic studies were launched. J-87, radioactively labeled with tritium, was traced through the systems of rodents and beagles to discover how and where it was absorbed, how much was absorbed, and how much was excreted. Although the doses were considerable, the animals quickly excreted it, and there were still no signs of toxicity.

Monkey studies unexpectedly settled the issue: the metabolic pathway J-87 took in monkeys was completely different, so instead of a multigram dose, only milligram doses were needed for good results.

Dr. Buchanan, the director of clinical research at Parke, Davis, refused to believe this. When the company's chief toxicologist confirmed that all the animal studies, including the latest monkey studies, were correct, Dr. Buchanan refused to believe him. A soft-spoken man who attributes his gray hair to J-87, Buchanan explained: "Now there were two concepts coming to the surface that were unacceptable. One, that a drug killed a virus."

His expression became incredulous. "And two, it didn't harm cells!" He resisted pressure, particularly from the toxicologist, to administer J-87 to people "to see what would happen." He ordered more animal studies instead.

After a review of all the J-87 animal experiments was published in 1968, Bob Buchanan received a telephone call from a professor at the University of Alabama, Charles Alford, who had read the report. "Alford said something ought to be done about this. He, too, had a null hypothesis—this can't possibly work. We agreed. But, we agreed also: we have to find out for sure."

Neither of them was about to propose that metabolic studies with radioactively labeled J-87 be conducted on normal people, so they settled on three cancer-patient volunteers. But how much to give? "Usually one is guided by one's animal studies," Buchanan said. "But our animal studies gave us no guidance whatsoever as to what the human dose should be." Even if the animal studies had indicated a predictable dosage based on body weight, no animal study is a fail-safe guide to human reaction. After repeated doses of penicillin, for instance, the normal intestinal bacteria of guinea pigs is replaced with a variety fatal to them.

"In determining the first human dosage of J-87, we had to take our knowledge and estimate—" Bob Buchanan stopped, tilting his head to one side. "What I mean is that our first dose was . . . pretty gutty." He and Alford took the dose least toxic in monkeys and divided it by ten. "We were too low. That amount didn't do much. Then we overshot. We got our dose a little bit high and found we were suppressing platelets. Then we came back down." Once the dosage was properly adjusted, patients seemed to tolerate J-87—beautifully.

"Then it was time to see if it worked," Buchanan continued. He went down to Birmingham, Alabama, carrying a small bottle of J-87 along in his pocket. He and Charles Alford injected it into a critically ill patient with a herpes skin infection. Within days the infection abated.

Encouraged, they administered J-87 to several more patients, and their infections also quickly disappeared. As far as the two researchers could tell, the compound didn't harm cells or have any ill effects on the immune system. But they recognized that they had not *proved* anything.

What they had was a small number of patients with herpes infections who *seemed* to get better after they were given injections of J-87. A full-fledged clinical study was in order, but there were hindrances. It would be risky and very costly, and not least of all, as far as Parke, Davis's board of directors was concerned, the potential return on their investment was slight.

The herpes viruses inflict myriad miseries. Herpes zoster, when it first attacks, causes chickenpox; the virus then retreats up the nerves enervating the skin, and hides out in the ganglia near the spinal column. Brought to life years later by unknown causes, the virus journeys down the nerves once more, causing the painful lesions of shingles. In patients whose immune systems have been impaired, such as cancer patients, this second attack can be fatal.

The herpes simplex viruses operate in a similar style, disappearing after an initial infection of the throat, eye, genitals, skin, or brain. If the virus is herpes simplex type 1 (HSV-1), the patient may experience a sore throat before the virus travels up the nerves from the mouth to the ganglion. There the virus rests until it is reactivated by sunlight, fever, stress, or other elements, and it voyages down the nerves to the lips, manifesting itself as a cold sore. The virus doesn't easily tire of this game and may play it over and over again.

Herpes viruses affecting the eye can impair vision or cause blindness. In the United States they were causing close to 20,000 cases of blindness a year. Other herpes viruses first infect the genitals. The infection often becomes latent, but if not, patients suffer frequent painful and distressing recurrences, and the virus, herpes simplex type 2 (HSV-2), can spread to a baby during the birth process with lethal consequences. The genital disease was becoming epidemic in the United States: millions suffered from it and hundreds of thousands more were infected every year.

Much rarer were herpes infections of the brain, or herpes encephalitis, an unusually deadly and macabre affliction. At first patients seemed merely confused, as if they had a viral illness of the nervous system. But the infection advanced with terrible speed, and within days they lapsed into coma. In a month, most victims were dead. Those who survived suffered such extreme neurologic damage that they were left permanently incapacitated, physically and mentally.

Parke, Davis had proceeded with trials of J-87 as an ointment for herpes eye infections, with outstanding results. The first trials had gone aground—when all six patients who received J-87 got worse—but it was found that the ointment's base was an eye irritant. A new base was substituted, and J-87 dramatically proved itself. It was not only far more effective than idoxuridine, it was nontoxic to cells.

But the company was not willing to invest in trials of J-87 for internal use. The chances for success were slight to begin with, and even if it worked, Parke, Davis could not patent the product. Even if they were able to patent their method of making J-87, the cost of the necessarily lengthy human trials would not be earned back for years, if ever.

Luckily, a telephone call solved the problem. "Did George call me or did I call him?" Bob Buchanan said. "I think he called me." Dr. George Galasso, chief of the National Institutes of Health's Antiviral Substances Program, had been organizing studies of promising new drugs. Interferon was at the top of his list, and interferon studies supported by NIH were already in progress. Of the several other experimental antiviral compounds that existed by 1969, the one that seemed to Galasso and his advisors to have the most promise—on the basis of Parke, Davis's work—was J-87. The government, Galasso told Buchanan, was ready to fund human trials of J-87 for internal use.

In view of the unknown risk of toxic side effects from any antiviral agent, it was resolved that studies would be done on only the most serious herpes diseases: herpes encephalitis, herpes zoster in patients with impaired immune systems, and herpes infections in newborns. The grant for testing J-87 against herpes encephalitis was awarded to Dr. Alford at the University of Alabama.

While herpes is among the most common causes of fatal encephalitis in the United States, brain inflammation can be caused by many other viruses and bacteria. Previous studies of idoxuridine had neglected to establish that patients treated with the compound had herpes encephalitis in the first place. So this time absolute proof in the form of a brain biopsy—a risky procedure—was required.

Even aside from this requirement, finding patients for the study was difficult. The disease is rare, it runs its course rapidly, and a diagnosis often is not made until the late stages. Further, as neurosurgeons are reluctant to perform a brain biopsy on patients who are not gravely ill, those who were eventually referred to the study tended to be the sickest—some had already suffered irreversible neurologic damage—and hence the most difficult to treat.

But late in 1972 the study, which was directed by Dr. Alford and his associate Dr. Richard Whitley and involved fifteen medical centers across the country, began. It was as scientifically air-tight as Alford, Whitley, and George Galasso could make it: a double-blind, placebo-controlled trial.

Patients with herpes encephalitis were chosen at random to receive either J-87, idoxuridine, or a harmless, inert substance, a placebo. The infusions, given intravenously, were coded, so neither physicians nor patients knew who was receiving what. After ten days of treatment, the patients were monitored for signs of the virus and toxic effects, and the survivors were observed for two years after that.

As time passed it became clear that some of the patients were faring very poorly compared to the others, and after two years the doctors felt compelled to break the code on those patients to find out what had gone wrong. All of them had received idoxuridine, which was being used at various medical centers by then to treat herpes encephalitis. Following their report on this finding, it was barred from further internal use.

When the code was completely broken, it was learned that most of the surviving patients had received J-87. A furious debate ensued. Ethics committees at some of the institutions participating in the study insisted that because the drug worked, it had to be given to anyone who might be helped and that to withhold it from the fatally ill was unethical.

"But that was not our way," George Galasso explained. "We maintain that if you have a drug that works, you may have a drug that can also cause some damage." He argued that without continuing a placebo control, many more patients would have to be treated to prove the usefulness of the drug. With a disease as rare as herpes encephalitis, that could take decades. Even with a placebo-controlled study, "statistical significance" had not yet been achieved.

The study continued with a placebo control and J-87 for two more years, and when five more patients had been treated, making a total of twenty-eight cases, the army of statisticians enlisted in the study was satisfied.

J-87 worked. A drug had combatted a lethal viral disease for the first time in history!

On August 11, 1977, the announcement was made in the *New England Journal of Medicine*: adenine arabinoside—J-87— significantly reduced the death rate of herpes encephalitis, from 70 to 28 percent. When the severity of the cases treated was taken into account, the actual reduction was even greater. More than half the treated survivors suffered little or no neurologic damage, and the drug had no serious toxic side effects.

The results seemed incredible, even to those who had directed the work, and they awaited the results of further J-87 studies with trepidation.

Additional trials confirmed that J-87 worked: in infants born with herpes infections, the death rate was reduced by a similar ratio, from 75 to 30 percent. The drug was also effective against shingles, and a continuation of the herpes encephalitis study upheld the original results.

A little more than a year later, in October 1978, adenine arabinoside, or ara-A, was approved by the Food and Drug Administration for internal use.

All those concerned with the work agreed that a most unusual collaboration had produced ara-A. The public and private sectors and the academic community had all worked together and, not least unusual of all, achieved a common goal. Investigators at the pharmaceutical firm first discovered the compound was useful in humans. The National Institutes of Health determined that the study should be done and funded it. And two University of Alabama professors designed and directed the study that proved ara-A worked.

Henry Dion unscrewed the lid of a brown glass jar and shook out into his palm a small heap of dazzling white powder. He gazed fondly at it. "It may sound corny," Dion said. "But for a chemist, when you hold something in your hand that no one else has—or at least you think you're the only one who has it—that's worth everything. But there's no use discovering a new drug," he added, "if you can't make it, and the early cultures were yielding only five to ten micrograms of ara-A. That's *nothing!*"

Although it was easier to ferment than to synthesize ara-A, developing the fermentation process involved years of trial and error. The organism itself was critical. "We had some strains of the organism that would make a lot of ara-A in a shake-flask, but wouldn't make any in a stirred-jar," Dion said. "Other organisms would make it in a small stirred-jar, but not in large tanks."

Once they found a cooperative, productive strain of the ara-A-producing organism, *Streptomyces antibioticus,* they could not seem to get rid of it. "One of the creatures they developed in microbiology was such a *slimy* creature you couldn't get it *out* of the beer," Dion said. "We told them, 'Hey, you're going down the wrong path with this thing' and sent it back to them. There's a lot of interplay between the microbiologists and the chemists," he said, chuckling.

Then once the filtration problem was solved, they couldn't isolate the pure compound from the clarified beer in sufficient quantities. The chemist at the next bench complained to Henry Dion about it. Dion ruminated and hit on the solution: "I said to myself, hey, there's one property that sticks out like a sore thumb. You use that to isolate it." The trait? Dion hesitated. "It's a bad trait of the drug. I don't like to mention it." But immediately he relented—it was a good story.

"Solubility. Ara-A's not very soluble in water. Under certain conditions it will precipitate out. We reduced the volume of the beer, chilled it, and it just plopped out!" This enabled the company to start producing ara-A by the pound and to do the exhaustive testing necessary to develop a new drug.

A few steps from Henry Dion's office in an old brick building at the company's Detroit headquarters, 5,000 gallons of gazpacho-hued soup swirl about in an immense green tank. Ordinarily, the fermentation process might be contracted out, but then the secret would also be out, and because Parke, Davis does not hold a patent on the product itself, the recipe for ara-A is jealously guarded. The actual processing is kept literally under lock and key close by the executive offices. Everyone in the building, from Dion to elevator operators to vice-presidents, wears a plastic identification card on a chain around the neck.

Into the green tank, Dion acknowledged, go glucose, soybean meal, water, minerals, salts, gallons of an antifoam agent, and many other ingredients. . . . He became vague—maybe corn syrup, maybe black strap molasses, maybe malt liquors, maybe corn meal, or fish meal, peanut meal, or oatmeal. He hinted at a special ingredient not ordinarily used in the pharmaceutical business, but said that because of labor costs all the ingredients are readily available and cheap. Even so, it costs $1,000 just to fill the tank, just to start the process.

The concoction is sterilized and allowed to cool before 125 gallons of "seed culture" are dumped in. Then the lid is closed tightly and the fermentation begins. For five days, under vigilantly controlled conditions, *Streptomyces antibioticus* obligingly makes ara-A. Filtered air is continually bubbled up from the bottom of the tank, and as some of the nutrients are used up, the fermentation is "fed." Once arabinose was added with the thought that it might increase the yield, but it did not.

By altering temperature, aeration, the media, and the proportions of each ingredient, the microbiologists eventually devised a cheap, foolproof method of producing ara-A. "These little exercises," Dion remarked, "cost a lot of money." All told, the company estimates it spent $5 million developing ara-A, and the National Institutes of Health spent in addition $2 million.

The company has yet to reap any profits from J-87, nor does it expect to—at least not directly. While ara-A can cure some of the worst herpes infections—encephalitis, newborn infections, and severe shingles—it is not effective against others. The developer of a cure for herpes genital infections would have a big moneymaker, though so far no one has been able to come up with one. "In the laboratory, ara-A is active against herpes vaginitis," Henry

Dion said, "but we can't get the right formulation. They're quite different—laboratory animal studies and human use." A remedy for cold sores would also be very profitable, but because ara-A does not eradicate the virus in its latent state, it does not completely cure the condition.

Ara-A remains a paradox. In cultures, it fractures the chromosomes of human leukocytes, but in the human body it doesn't. As Parke, Davis's chief toxicologist marveled, "In clinical studies, ara-A was shown to be extraordinarily kind to bone marrow and lymphoid tissue, a feature totally out of context with a drug that breaks chromosomes in such a grand style."

Numerous theories have been advanced as to the secret of the wondrous white crystalline powder, but they all amount to the same thing: as ara-A is metabolized, either it or its phosphate byproducts inhibit one or more of the enzymes necessary to viral DNA synthesis, while the equivalent host cell enzymes, which are apparently less sensitive to it, remain unaffected.

Fortunately, understanding how antiviral compounds work has not been necessary to their development, and there are several in addition to ara-A in use. Amantadine prevents and treats influenza; however, its benefits may be limited on account of the notorious speed with which influenza spreads. Another is methisazone, which prevents smallpox, a disease that was wiped out, ironically, just as the compound was discovered—200 years after Edward Jenner discovered how to do it. There is also interferon, a protein that occurs naturally in humans and animals and is active against many viruses. The trouble is that interferon has "looked promising" for more than twenty years. Now it seems that interferon will not be the great panacea it was once hoped. It may help fight viral infections and cancer, but by itself it does not seem equal to the job. Lately, new, more potent compounds than ara-A have been developed, and one of them seems to be helpful in treating herpes genital infections. Certainly it won't be long before other, still better such magic bullets are found.

But ara-A, the pure white crystalline powder beneficently

manufactured by bacteria, unearthed by Cicillo Aldo, and put to use through the efforts of hundreds of chemists, microbiologists, technicians, physicians, and the most patient of patients, deserves to be remembered. It was the beginning.

Notes

I HAVE NOT INCLUDED FOOTNOTES with the thought that they serve to annoy readers more than anything else. This work is intended to be strictly accurate. The facts have been assembled and interpreted, of course, but every idea, opinion, and sentiment attributed to someone is based on facts. These facts were drawn from interviews, conversations, and correspondence; personal diaries, letters, and other papers of the protagonists; as well as standard sources. For any who may be curious, I list them here.

"Obscure, If Not Positively Unnatural"

Benenson, Abram S. "Smallpox." In Evans, Alfred S., ed., *Viral Infections of Humans: Epidemiology and Control.* New York: Plenum, 1976.

Davis, Dorland J. "Measurements of the Prevalence of Viral Infections." In Merigan, Thomas C., ed., *Antivirals with Clinical Potential*. Chicago: University of Chicago Press, 1976.

Kaplan, Martin M., and Webster, Robert G. "The Epidemiology of Influenza." *Scientific American* December 1977, 237: 88–106.

Locke, David M. *Viruses: The Smallest Enemy*. New York: Crown, 1974.

McNeill, William H. *Plagues and Peoples*. New York: Anchor/ Doubleday, 1976.

U.S. Department of Health, Education, and Welfare. "Common Cold." *Infectious Diseases Research Series*. Washington: June 1978.

The Sting of Death

Bordley, [John] B. [Beale]. *Yellow Fever*. Philadelphia, 1794(?).

Downs, Wilbur G. "Arboviruses." In Evans, Alfred S., ed., *Viral Infections of Humans: Epidemiology and Control*. New York: Plenum, 1976.

Lloyd, Wray; Theiler, Max; and Ricci, Nelda I. "Modification of the Virulence of Yellow Fever Virus by Cultivation in Tissues *in Vitro*." London: *Transactions of the Royal Society of Tropical Medicine and Hygiene* 29: 481–529, 1936.

McNeill, William H. *Plagues and Peoples*. New York: Anchor/ Doubleday, 1976.

Mathis, Constant. "*Le docteur Max Theiler, prix Nobel de Médecine 1951 pour ses travaux sur la fièvre jaune.*" Paris: *Bulletin de l'Académie Nationale de Médecine* 135: 562–64, 1951.

Powell, J.H. *Bring Out Your Dead: The Great Plague of Yellow Fever in Philadelphia in 1793*. Philadelphia: University of Pennsylvania Press, 1949.

Rush, Benjamin. *An Account of the Bilious Remitting Yellow Fever, As It Appeared in the City of Philadelphia, in the Year 1793*. Philadelphia: Thomas Dobson, 1794.

Smith, Hugh H., and Theiler, Max. "The Adaption of Unmodified Strains of Yellow Fever Virus to Cultivation *in Vitro*." *Journal of Experimental Medicine* 6: 801–808, 1937.

Strode, George K., ed. *Yellow Fever*. New York: McGraw-Hill, 1951.

Theiler, Max. "The Development of Vaccines Against Yellow Fever: Nobel Lecture." In *Nobel Lectures: Physiology or Medicine, 1942–1962*. New York: Elsevier, 1964.

———. Interviewed by Dr. Harriet Zuckerman, 1963 (unpublished).

———. Personal letters, 1936–1937 (unpublished).

———. "Max Theiler's Personal Recollections, 1901–1919" (unpublished).

Theiler, Max, and Smith, Hugh H. "The Effect of Prolonged Cultivation *in Vitro* upon the Pathogenicity of Yellow Fever Virus." *Journal of Experimental Medicine* 6: 767–86, 1937.

———. "The Use of Yellow Fever Virus Modified by *in Vitro* Cultivation for Human Immunization." *Journal of Experimental Medicine* 6: 787–800, 1937.

The Sculptor

Benison, Saul, ed. *Tom Rivers: Reflections on a Life in Medicine and Science.* Cambridge: M.I.T. Press, 1967.

Horstmann, Dorothy M. "Immunization Against Viral Infections." In Rothschild, Henry; Allison, Fred; and Howe, Calderon, eds., *Human Diseases Caused by Viruses: Recent Developments.* New York: Oxford University Press, 1978.

Melnick, Joseph L. "Enteroviruses." In Evans, Alfred S., ed., *Viral Infections of Humans: Epidemiology and Control.* New York: Plenum, 1976.

Paul, John R. *A History of Poliomyelitis.* New Haven: Yale University Press, 1971.

Sabin, Albert A. "Oral Poliovirus Vaccine: History of Its Development and Prospects for Eradication of Poliomyelitis." *Journal of the American Medical Association* 194: 872–76, 1965.

Anatomy of a Cold

Andrewes, Christopher H. "Fifty Years with Viruses." In Starr, Mortimer P.; Ingraham, John L.; and Raffel, Sidney, eds., *Annual Review of Microbiology.* Palo Alto: Annual Reviews, 1973.

———. *In Pursuit of the Common Cold.* London: Heinemann, 1973.

———. *The Natural History of Viruses.* London: Weidenfeld and Nicolson, 1967.

———. "Recollections." In Melnick, Joseph L., ed., *International Virology I.* New York: S. Karger, 1969.

Cate, Thomas R. "Rhinoviruses." In Knight, Vernon, ed., *Viral and Mycoplasmal Infections of the Respiratory Tract.* Philadelphia: Lea and Febiger, 1973.

Davenport, Fred M. "Influenza." In Evans, Alfred S., ed., *Viral Infections of Humans: Epidemiology and Control.* New York: Plenum, 1976.

Gwaltney, Jack M. "Rhinoviruses." In Evans, Alfred S., ed., *Viral Infections of Humans: Epidemiology and Control.* New York: Plenum, 1976.

Gwaltney, Jack M., and Hendley, J. Owen. "One If by Air, Two If by Hand." *American Journal of Epidemiology* 107: 357–61, 1978.

Gwaltney, Jack M.; Moskalski, Patricia B.; and Hendley, J. Owen. "Hand-to-Hand Transmission of Rhinovirus Colds." *Annals of Internal Medicine* 88: 463–67, 1978.

Jackson, George G., and Muldoon, Robert L. *Viruses Causing Common Respiratory Infections in Man.* Chicago: University of Chicago Press, 1975.

Locke, David M. *The Smallest Enemy.* New York: Crown, 1974.

Rothschild, Henry; Allison, Fred; and Howe, Calderon, eds. *Human Diseases Caused by Viruses: Recent Developments.* New York: Oxford University Press, 1978.

Stuart-Harris, Charles H., and Schild, Geoffrey C. *Influenza: The Viruses and the Disease.* Littleton, Mass.: Publishing Sciences Group, 1976.

Tyrrell, D.A.J. "In Honor of C.H.A. or Sir Christopher Howard Andrewes." In Pollard, Morris, ed., *Antiviral Mechanisms: Perspectives in Virology IX.* New York: Academic Press, 1975.

U.S. Department of Health, Education, and Welfare. "Acute Conditions, Incidence, and Associated Disability, United States, July 1977–June 1978." *Vital and Health Statistics Series 10,* no. 132. Washington: 1979.

U.S. Department of Health, Education, and Welfare. "The Common Cold: Relief But No Cure." *FDA Consumer.* Washington: September 1976.

The Virus That Ate Cannibals

Brody, Jacob A., and Gibbs, Clarence J. "Chronic Neurological Diseases: Subacute Sclerosing Panencephalitis, Progressive Multifocal Leukoencephalopathy, Kuru, Creutzfeldt-Jakob Disease." In Evans, Alfred S., ed., *Viral Infections of Humans: Epidemiology and Control.* New York: Plenum, 1976.

Gajdusek, D. Carleton. *Kuru: Collected Papers.* Bethesda: National Institutes of Health, 1969.

———. "Kuru in the New Guinea Highlands." In Spillane, J.D., ed., *Tropical Neurology.* New York: Oxford University Press, 1973.

———. *New Guinea Journals, 1957–1962.* Bethesda: National Institutes of Health, 1963–1968.

———. "Unconventional Viruses and the Origin and Disappearance of Kuru: Nobel Lecture." In *Les Prix Nobel: En 1976.* Stockholm: P.A. Norstedt and Söner, 1977.

Gajdusek, D. Carleton, ed. *Correspondence on the Discovery and Original Investigations on Kuru: Smadel-Gajdusek Correspondence, 1955–1958.* Bethesda: National Institutes of Health, 1975.

Gajdusek, D. Carleton, and Zigas, Vin. "Degenerative Disease of the Central Nervous System in New Guinea: The Endemic Occurrence of Kuru in the Native Population." *New England Journal of Medicine* 257: 974–78, 1957.

Gajdusek, D. Carleton; Gibbs, Clarence J.; and Alpers, Michael, eds. *Slow, Latent and Temperate Virus Infections.* Bethesda: National Institutes of Health, 1965.

A Perfect Crime

Baltimore, David. "Retroviruses and Cancer." *Hospital Practice* 13: 49–57, 1978.

———. "The Strategy of RNA Viruses." In *The Harvey Lectures, 1974–1975.* New York: Academic Press, 1976.

———. "Viral RNA-dependent DNA Polymerase." *Nature* 226: 1209–11, 1970.

———. "Viruses, Polymerases, and Cancer: Nobel Lecture." In *Les Prix Nobel: En 1975.* Stockholm: P.A. Norstedt and Söner, 1975.

Baltimore, David; Huang, Alice S.; and Stampfer, Martha. "Ribonucleic Acid Synthesis of Vesicular Stomatitis Virus, II: An RNA Polymerase in the Virion." *Proceedings of the National Academy of Sciences* 66: 572–76, 1970.

Dulbecco, Renato. "From the Molecular Biology of Oncogenic DNA Viruses to Cancer: Nobel Lecture." In *Les Prix Nobel: En 1975.* Stockholm: P.A. Norstedt and Söner, 1975.

Miller, George. "Epidemiology of Burkitt Lymphoma." In Evans,

Alfred S., ed., *Viral Infections of Humans: Epidemiology and Control.* New York: Plenum, 1976.

Nahmias, Andrew J., and Josey, William E. "Epidemiology of Herpes Simplex Viruses 1 and 2." In Evans, Alfred S., ed., *Viral Infections of Humans: Epidemiology and Control.* New York: Plenum, 1976.

Nahmias, Andrew J.; Josey, William E.; and Oleske, James M. "Epidemiology of Cervical Cancer." In Evans, Alfred S., ed., *Viral Infections of Humans: Epidemiology and Control.* New York: Plenum, 1976.

Rothschild, Henry; Allison, Fred; and Howe, Calderon, eds. *Human Diseases Caused by Viruses: Recent Developments.* New York: Oxford University Press, 1978.

Temin, Howard M. "The DNA Provirus Hypothesis: The Establishment and Implications of RNA-directed DNA Synthesis: Nobel Lecture." In *Les Prix Nobel: En 1975.* Stockholm: P.A. Norstedt and Söner, 1975.

Temin, Howard M., and Mizutani, Satoshi. "RNA-dependent DNA Polymerase in Virions of Rous Sarcoma Virus." *Nature* 226: 1211–13, 1970.

Pay Dirt

Bergmann, Werner, and Feeney, Robert J. "Contributions to the Study of Marine Products, XXXII: The Nucleosides of Sponges, I." *Journal of Organic Chemistry* 16: 981–87, 1951.

Buchanan, Robert A., and Hess, Frank. "Vidarabine: Pharmacology and Clinical Experience." *Pharmacology and Therapeutics* 8: 143–71, 1980.

De Rudder, Jean, and Privat de Garilhe, Michel. "Inhibitory Effect of Some Nucleosides on the Growth of Various Human Viruses in Tissue Culture." In Hobby, Gladys L., ed., *Antimicrobial Agents and Chemotherapy, 1965.* Washington: American Society for Microbiology, 1966.

Galasso, George J.; Merigan, Thomas C.; and Buchanan, Robert A., eds. *Antiviral Agents and Viral Diseases of Man.* New York: Raven Press, 1979.

Gunby, Phil. "New Anti-Herpes Virus Drug Being Tested." *Journal of the American Medical Association* 243: 1315, 1980.

Hirsch, Martin S., and Swartz, Morton N. "Antiviral Agents." *New England Journal of Medicine* 302: 903–906, 949–53, 1980.

Jackson, George G. "Chemotherapy." In Rothschild, Henry; Allison, Fred; and Howe, Calderon, eds., *Human Diseases Caused by Viruses: Recent Developments.* New York: Oxford University Press, 1978.

Lauter, Carl B. "Herpes Simplex Encephalitis: A Great Clinical Challenge." *Annals of Internal Medicine* 93: 696–97, 1980.

Maugh, Thomas H. "Chemotherapy: Antiviral Agents Come of Age." *Science* 192: 128–32, 1976.

Pavan-Langston, Deborah; Buchanan, Robert A.; and Alford, Charles A., eds. *Adenine Arabinoside: An Antiviral Agent.* New York: Raven Press, 1975.

Pratt, William. *Chemotherapy of Infection.* New York: Oxford University Press, 1977.

Schabel, F.M. "The Antiviral Activity of 9-Beta-D-Arabinofuranosyladenine." *Chemotherapy* 13: 321–38, 1968.

Whitley, R.J.; Soong, S.-J.; Dolin, R.; Galasso, G.J.; Ch'ien, L.T.; Alford, C.A.; and Collaborative Study Group. "Adenine Arabinoside Therapy of Biopsy-Proved Herpes Simplex Encephalitis." *New England Journal of Medicine* 297: 289–94, 1977.

Index